Making ChatGPT Work for You

Getting the Most Out of Generative AI as a Non-Techie

Lydia Evelyn

Making ChatGPT Work for You: Getting the Most Out of Generative AI as a Non-Techie

Lydia Evelyn
Catalonia, Spain

ISBN-13 (pbk): 979-8-8688-1444-0   ISBN-13 (electronic): 979-8-8688-1445-7
https://doi.org/10.1007/979-8-8688-1445-7

Copyright © 2025 by Lydia Evelyn

Managing Director, Apress Media LLC: Welmoed Spahr
Acquisitions Editor: Shivangi Ramachandran
Development Editor: James Markham
Editorial Assistant: Jessica Vakili

Cover designed by eStudioCalamar

Distributed to the book trade worldwide by Springer Science+Business Media New York, 1 New York Plaza, New York, NY 10004. Phone 1-800-SPRINGER, fax (201) 348-4505, e-mail orders-ny@springer-sbm.com, or visit www.springeronline.com. Apress Media, LLC is a Delaware LLC and the sole member (owner) is Springer Science + Business Media Finance Inc (SSBM Finance Inc). SSBM Finance Inc is a **Delaware** corporation.

For information on translations, please e-mail booktranslations@springernature.com; for reprint, paperback, or audio rights, please e-mail bookpermissions@springernature.com.

Apress titles may be purchased in bulk for academic, corporate, or promotional use. eBook versions and licenses are also available for most titles. For more information, reference our Print and eBook Bulk Sales web page at http://www.apress.com/bulk-sales.

Any source code or other supplementary material referenced by the author in this book is available to readers on GitHub. For more detailed information, please visit https://www.apress.com/gp/services/source-code.

If disposing of this product, please recycle the paper

Table of Contents

TABLE OF CONTENTS

About the Author

Lydia Evelyn is the author of *Beginning ChatGPT for Python: Build Intelligent Applications with OpenAI APIs.* She is a proficient technical writer and a skilled Python developer with a professional background in ChatGPT. She has effectively utilized ChatGPT, coupled with prompt engineering, in multiple client projects within the realm of technical publishing. Lydia boasts extensive programming expertise in both Python and Java.

Introduction

Artificial intelligence is reshaping the way we work, create, and interact. From content generation to data analysis, AI-powered tools like ChatGPT are becoming indispensable for students, professionals, and everyday users alike. However, many people barely scratch the surface of what ChatGPT can actually do. If you have ever felt like you're not getting the most out of ChatGPT, or if you're stuck using generic prompts that don't quite give you what you need, this book is for you.

Making ChatGPT Work for You is a hands-on companion designed to help you unlock the full potential of this powerful AI tool. This isn't just a book on what ChatGPT can do—it's a practical guide that will show you exactly how to use it effectively. With over 100 curated prompts, you'll learn how to craft precise, results-driven instructions that allow you to generate high-quality content, analyze data, streamline your workflow, and even use ChatGPT for creative projects.

By the time you finish this book, you'll understand the art of prompt engineering—the key skill that separates casual users from AI power users. You will learn how to ask the right questions, refine ChatGPT's responses, and even teach the AI to write in different styles, generate personalized reports, and perform complex tasks like budget planning and research synthesis.

But we're not stopping at just text. With ChatGPT's DALL-E and Sora, AI-generated images and short videos are now within your grasp. Whether you're looking to visualize concepts for presentations, generate eye-catching graphics, or create video content for social media, this book will guide you through using these cutting-edge features.

Think of this as your prompt cookbook—a reference you can turn to whenever you need a jumpstart on an idea, help with a complex task, or an AI-powered assistant to make your life easier. Whether you're a student, writer, entrepreneur, teacher, or just curious regarding AI, you'll find practical and actionable ways to incorporate ChatGPT into your daily routine.

The future of AI-assisted work and creativity is already here—let's make sure you're ahead of the curve.

CHAPTER 1

Sorry, but You're (Probably) Not Using the Best Prompts to Use ChatGPT to Its Highest Potential

Since the inception of ChatGPT, it's been seen and touted as a miracle drug. If you believe all the hype, it can create an entire anthology of books from a single prompt! Don't believe the hype.

I personally had the opportunity to use it in multiple professional contexts, from marketing to engineering and from software development to database analysis. And for many business professionals, they're using ChatGPT **the wrong way**. For some, using ChatGPT for work gets frustrating when it doesn't produce the miraculous results they were expecting. For others, I've seen that users may not know how to properly fact-check results from ChatGPT. This results in work tasks that have embarrassingly obvious errors. That may even be why you picked up this book!

© Lydia Evelyn 2025
L. Evelyn, *Making ChatGPT Work for You*, https://doi.org/10.1007/979-8-8688-1445-7_1

Things You May (or May Not) Know That ChatGPT Can Do

Here's a list of some really cool things you can do with ChatGPT.

Real Interview Prep with Practice Questions

Did you know that you can use ChatGPT in order to prep yourself for a job interview? Yes, seriously. You can actually have a one-on-one **verbal** conversation with ChatGPT and have it to play the role of the job recruiter from the HR department of the company you want to work for. It will ask you relevant questions about the job, giving you the opportunity to practice and hone your responses for the real thing. Be sure to check out Chapter 8, which is all about chatting and conversing with ChatGPT for productivity.

Take the Pain Out of Reading Really Long Documents

Did you know that you can leverage ChatGPT to analyze your documents? What that means is if you have a document—be it a Word document, an Excel file, or even a PDF—you can send it to ChatGPT for analysis. You can then ask questions about the contents of the file itself. For some business professionals, this is a game-changer because you can take a marketing brief or a white paper and ask ChatGPT **anything** about the contents. For example, have you seen the terms and conditions of the apps that you use every day? You almost need a lawyer to read them for you before you hit that "Accept" button. Well, in Chapter 6, we do the next best thing—we're taking the Terms and Conditions for Facebook as a PDF file, and we're going to ask ChatGPT what they're *really* asking you to accept.

Create the Exact Image That You Need for Your Project

Did you know that you can generate really professional-looking photos, images, icons, and illustrations using ChatGPT? You can use these images for any purpose you want and share what you've made with marketing teams, graphic designers, or managers and communicate your ideas in a more visual way that helps get everyone on the same page. If you're curious about how to do this **well**, be sure to check out Chapter 13 where we go through all of the steps to crafting the right prompt for image generation.

Write Engaging Long-Form Content with the Canvas Feature

Of course, you know that you can write an article with the help of ChatGPT, but did you know that you can use ChatGPT's Canvas feature to take it to another level? This incredible feature allows you to be able to write with ChatGPT side by side, in a collaborative window called a "Canvas" that helps you make changes and critiques with ChatGPT. Think of it as your very own collaboration partner for writing articles, documents, or even book chapters. Be sure to check out Chapter 5 to see this in action!

Ensure You're Getting Factual and Accurate Results from ChatGPT

Everyone knows that ChatGPT can provide mistaken results. But did you know that you can circumvent this issue entirely simply by making ChatGPT cite its sources? ChatGPT can search the Internet and provide relevant links when requested, allowing you to double-check the information yourself. Not only is this a handy feature that allows you to ensure your information comes from reliable sources, we consider this to be one of the best practices of using ChatGPT. Find out more about all of the best practices of using ChatGPT to get accurate results in Chapters 3 and 4.

All of this is just the beginning. ChatGPT is a powerful tool that can do some amazing things when used properly. Now having that stated, who are you, dear reader?

Who Is This Book For?

This book is for *anyone* who wants to use ChatGPT for *any* reason— whether it's boosting productivity at work, saving time and energy for content creation, or even saving yourself the brainpower on a daily basis by letting a chatbot plan and suggest things for you. This book is the only book you'll ever need to coax ChatGPT into making your life much easier by creating, dissecting, sorting, and understanding information in ways that can accommodate any use case you can imagine. More than that, we're going to explore some little-known features that have a great impact on your productivity.

Note   You will need a subscription to ChatGPT Plus to follow most of the tutorials in this book. While the concepts of prompt engineering will apply across the board, many specific features that will be covered in this book will require a subscription to ChatGPT Plus to access. If you don't know how to subscribe, an explanation on the process will be provided later in this chapter.

To be perfectly clear, you don't need to know anything about AI or chatbots to get started with this book. In this chapter, we're even going to walk you through the process of signing up for a ChatGPT account. This book will guide you with step-by-step tutorials and teach you prompt engineering techniques (if you've never heard that term before, stay tuned—Chapter 2 has you covered). You'll learn how to use ChatGPT for various tasks and projects, discover abilities of ChatGPT you never knew existed, and even generate images. We're also going to address the elephant in the room. ChatGPT will never take your job. Instead, knowing how to use it properly will make you more productive and creative so you can work on projects and ideas efficiently.

But before we dive into all that, let's start with the basics.

Let's Talk About ChatGPT: What Is It?

I'm pretty sure that everyone knows that AI stands for "artificial intelligence," but how does it actually work? In the simplest of terms, when a software application needs to make a decision **without being instructed to do so**, it's using some form of artificial intelligence, or an algorithm. Let's talk about what's going on behind the scenes when you hit "like" on a TikTok video. Essentially, besides indicating to the creator and others that you enjoyed what you saw, you're also inadvertently telling TikTok to find another video just like it. Additionally, you're also informing TikTok various other things, such as

- How long did you watch the video?

- Did you watch it multiple times?

- What hashtags were used?

TikTok's algorithm takes all this information, processes it, and then suggests videos it thinks you'll enjoy. It also looks at what other people with similar interests liked. This is the "intelligence" part of "AI"—it's software that uses lots of data to make suggestions or decisions.

The same principle applies for applications across the board. Spotify's recommendations are based upon the songs you play, how often you play them, and what other listeners with similar tastes enjoy. Spell-check works similarly by considering common typos, sentence structure, and context to correct your writing.

So, AI isn't a robot behind the scenes doing everything for you. It's smart software that learns from data and makes educated guesses to help you out. By breaking it down this way, we can see that AI is all about using data to make smart decisions.

What ChatGPT Can Do (and What It Can't Do)

AI is everywhere, but it's only recently that it's become a hot topic outside the tech world. This is largely due to ChatGPT's breakout success with GPT 3 in late 2022. Since then, the technology has only continued to advance. Some are thrilled and excited about ChatGPT, while others, particularly in the workforce, find the software intimidating and worry that AI might replace their jobs. After all, we're seeing AI create images that look like complex artwork, songs that sound like they're sung by real singers, and articles and stories that seem to have been written in minutes by a piece of software—sometimes even quicker than that.

However, it's not as simple as asking ChatGPT to create a masterpiece like Da Vinci or the next great American novel with the press of a button. Creating something amazing with ChatGPT requires thoughtful input that leverages its massive collection of data to produce the desired result.

In other words, ChatGPT is only as creative as the person using it. How does all of this work? Well, it all starts with a prompt.

So, What's a Prompt?

Simply put, a prompt is the input text you give to ChatGPT. That's it. You don't have to be a tech whiz in order to use ChatGPT competently. It's actually very easy to get ChatGPT to do something as simple as creating a vegan meal plan for the week ahead or as complex as planning a mailing list campaign for a marketing team. The key to getting useful results is knowing how to give ChatGPT the right input in order to receive an output that fits your needs.

Now that we know what ChatGPT is, how to use it, and what a prompt is, let's create an account with OpenAI, the company that owns ChatGPT.

We would—except... we want to cover one more thing first.

What Does Nondeterminism Mean, and What Does It Have to Do with ChatGPT?

When you use ChatGPT, you'll notice that if you ask the same question multiple times, you get slightly different answers each time. This is actually by design. ChatGPT can generate a variety of responses, and this ability to respond differently to the same prompt multiple times is called nondeterminism.

Why Is Nondeterminism Important?

Imagine talking to a friend who gives you the exact same response every time you ask a question. It would feel very unnatural and robotic, right? Nondeterminism helps ChatGPT seem more like a real person, making conversations more natural and engaging.

Besides sounding more human, nondeterminism makes ChatGPT more useful. Sometimes there are several good ways to answer a question or respond to a prompt. When you're using ChatGPT for productivity, nondeterminism helps you explore different possibilities and gives you more options if you don't get the answer you're looking for the first time.

Listing 1-1 shows an example of what this looks like.

Listing 1-1. Even if you barrage ChatGPT with the same question, you'll get a different response every time

```
User: "What's a good way to spend a weekend?"
ChatGPT: "You could go hiking in a nearby park."
User: "What's a good way to spend a weekend?"
ChatGPT: "How about visiting a museum or going to a movie?"
User: "What's a good way to spend a weekend?"
ChatGPT: "You might enjoy having a relaxing day at home with a
good book."
```

All these answers are good, but they're different. This variety makes ChatGPT more helpful and fun to use because it offers diverse suggestions rather than repeating the same thing over and over.

Alright, now for real this time. Let's create an account with OpenAI and start using ChatGPT for ourselves.

Let's Get Started: How to Create an Account with ChatGPT

Now that we've covered the basics of what ChatGPT is and how to use it, let's get you set up with your own account. Creating an account with OpenAI is simple and straightforward. Follow these steps, and you'll be ready to start using ChatGPT.

In order to create an account, simply go to `https://chatgpt.com/` and sign up, as shown in Figure 1-1.

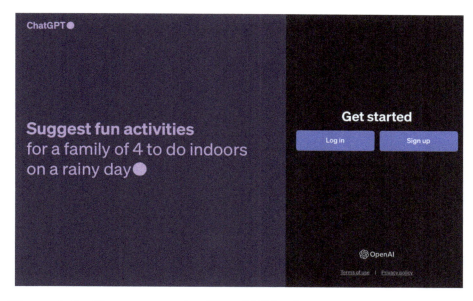

Figure 1-1. *Sign-up page for ChatGPT by OpenAI*

The process is as simple as can be. Click "Sign up" if you don't already have an account, and you'll be taken to a screen that gives you multiple sign-up options, as shown in Figure 1-2.

Create an account

Email address

Continue

Already have an account? Login

—— OR ——

G   Continue with Google

▦   Continue with Microsoft Account

   Continue with Apple

Terms of Use | Privacy Policy

Figure 1-2. *You can sign up for ChatGPT by creating a new account*
with your email address or using your existing account with Google,
Microsoft, or Apple

Understanding the Anatomy of a Chat Window with ChatGPT

After signing up, you'll be taken straight to your chat window to interact with ChatGPT. If you notice, the site gives you some recommendations for conversations to start with. Figure 1-3 displays the chat window that appears when you're signed in to the ChatGPT site.

Figure 1-3. *The chat window for using ChatGPT—a numbered guide*

Let's look at what each annotated part of the page can actually do:

1. This shows the model of ChatGPT you're currently using. For users that haven't upgraded to ChatGPT Plus, you cannot switch between ChatGPT models, and behind the scenes, the default is ChatGPT 4o mini. Upgrading, however, allows you to switch between ChatGPT 4o mini and all of the other models, such as 4o canvas, o1 mini, and o1, each version of greater quality than the last.

11

Note   For the rest of this book, unless otherwise noted, we're going to be using ChatGPT 4o, an option you can select from the chat window here.

2. This is the text box that allows you to input the text to send to ChatGPT as a prompt.

3. This is the send button; however, alternatively, you can hit "enter" on your keyboard to send a prompt.

4. This is where the history of your chats will be stored. You can go back to the very first conversation you've ever had with ChatGPT from here!

5. You can create a new chat conversation with this button.

6. Clicking this icon allows you to hide the history panel.

7. This is your profile icon, and clicking it allows you to manage your settings, such as the theme color, the default language, and more.

Because of the fact that we're using the latest and most powerful features ChatGPT has to offer, we're going to upgrade to ChatGPT Plus. It's a simple process, but the next step will illustrate exactly what you need to do.

Upgrading to ChatGPT Plus to Get All the Latest Features

If you're on the chat window on the ChatGPT site, you can click the **Upgrade Plan** button in the lower-left part of the screen. Afterward, you just select the plan you prefer and sign up. Figure 1-4 shows what that looks like.

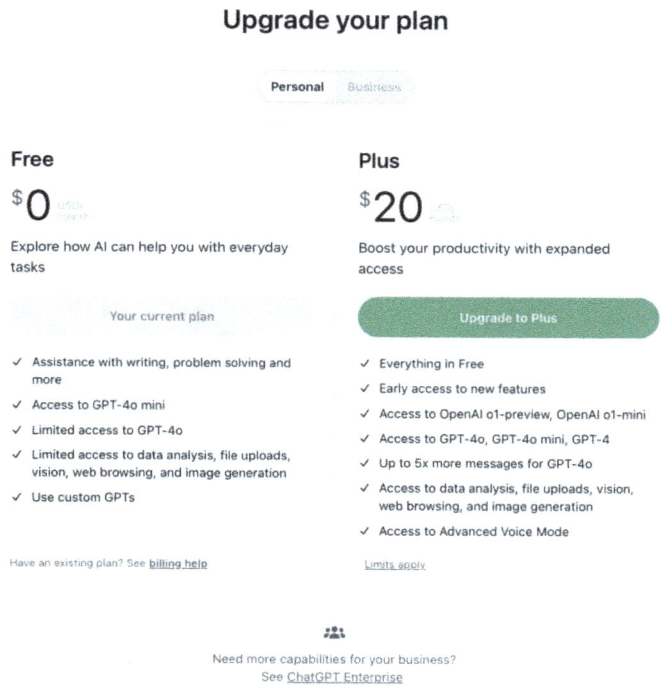

Upgrade your plan

Personal   Business

Free

$0

Explore how AI can help you with everyday tasks

Your current plan

✓ Assistance with writing, problem solving and more

✓ Access to GPT-4o mini

✓ Limited access to GPT-4o

✓ Limited access to data analysis, file uploads, vision, web browsing, and image generation

✓ Use custom GPTs

Have an existing plan? See billing help

Plus

$20

Boost your productivity with expanded access

Upgrade to Plus

✓ Everything in Free

✓ Early access to new features

✓ Access to OpenAI o1-preview, OpenAI o1-mini

✓ Access to GPT-4o, GPT-4o mini, GPT-4

✓ Up to 5x more messages for GPT-4o

✓ Access to data analysis, file uploads, vision, web browsing, and image generation

✓ Access to Advanced Voice Mode

Limits apply

Need more capabilities for your business?
See ChatGPT Enterprise

Figure 1-4. *The upgrade page shows all the benefits available with ChatGPT Plus*

Conclusion

In this chapter, you learned what ChatGPT is and why it's nondeterministic. You also learned how to create an account with ChatGPT and upgrade to ChatGPT Plus. In the next chapter, we'll look more closely at prompts and see how to illicit the best responses.

CHAPTER 2

Prompt Engineering: The Key to Being a Pro at Using ChatGPT

When you set out to use ChatGPT to help you with a project, come up with an idea, or streamline your productivity at work, there's one thing that will set you apart from a very dry and unhelpful response from ChatGPT to something unique and usable for any number of use cases—good prompt engineering. In this chapter, we're going to take a closer look at prompt engineering and review some of the most popular conventions for getting helpful responses.

Explaining Prompt Engineering by Using Prompt Engineering

Let's kick things off with an example. Prompt 2-1 is our very first prompt. We're going to ask ChatGPT to explain what prompt engineering is all about. This will give us a concrete idea of how to interact with ChatGPT and get useful information from it.

© Lydia Evelyn 2025
L. Evelyn, *Making ChatGPT Work for You*, https://doi.org/10.1007/979-8-8688-1445-7_2

Prompt 2-1. The beginning! What's prompt engineering?

```
In simple terms, explain what prompt engineering is.
```

This is a nice, simple prompt. Unlike a Google search, where we might just ask "What is prompt engineering?" we're guiding ChatGPT toward the type of response we want and telling it how we want it. We're not looking for a lengthy article—just a straightforward answer to a direct question. Let's see how ChatGPT answered in this example, shown in Response 2-1.

Response 2-1. A simple introduction to prompt engineering

```
Prompt engineering is the process of crafting and refining the
questions or instructions (prompts) that are given to an AI to
generate useful or creative responses. It involves designing
the prompt in a way that guides the AI to understand and
produce the desired outcome more effectively. This can include
specifying details, adjusting the tone, or focusing the request
to leverage the AI's capabilities optimally. The goal is to
enhance the AI's performance by providing clear, specific, and
well-thought-out prompts.
```

This is a decent explanation of prompt engineering that helps us understand it in simple terms, and it's certainly easier to read here compared to reading an entire article on the subject. However, in my opinion, it's a bit too bare and maybe a little complex and dry. So let's use prompt engineering to get a better response.

Prompt 2-2. Clarifying our needs to ChatGPT

```
Give me a more conversational answer and go into more detail so
I can learn best practices for prompt engineering.
```

What are we looking for in this response? We want ChatGPT to explain this in a way that anyone can understand, using simpler terms and a bit more detail. We want an explanation that not only defines prompt engineering but also gives us practical tips on how to use it to get the best results. We've updated our input, so let's look at the results in Response 2-2.

Response 2-2. Getting a better response from ChatGPT on how to get good responses from ChatGPT (talk about inception)

Sure, I'd be happy to dive deeper into prompt engineering! It's a bit like being a director in a play—you have to give clear and thoughtful instructions to get the best performance out of your actors, or in this case, the AI.

When you're prompt engineering, you're basically fine-tuning the way you ask the AI questions or give it tasks. The clearer and more specific your instructions, the better the AI can perform. Here are a few best practices:

1. Be Specific: Ambiguity can lead the AI down the wrong path. If you need a specific type of answer, make sure your prompt reflects that. For example, if you want a detailed explanation, specify that instead of just asking a broad question.

2. Provide Context: Sometimes, the AI needs a bit of background to give you the right answer. If your question relates to a specific topic or requires certain knowledge, include that context in your prompt.

3. Use Clear Language: Avoid jargon, unless it's relevant to the query, and try to be as clear as possible. The AI is good at following instructions but can get tripped up by overly complex language or double negatives.

4. Iterative Refinement: Don't be afraid to refine your prompts based on the responses you get. Sometimes, the first response from an AI gives you a better idea of how to ask a more effective follow-up question.

5. Experiment: Try different phrasings or approaches to see what yields the best results. You might find that slight changes in how you phrase a question can make a big difference in the type of answers you receive.

Think of it as a collaborative effort where your role is to guide the AI with precision and clarity, enhancing the likelihood of getting back exactly what you need.

This is excellent! With a little more input, we were able to get ChatGPT to give us very helpful instructions on how to... well, use ChatGPT.

This is a great basis to start with, and to add to it, we're going to take a look at the most common approaches to prompting ChatGPT for various use cases.

Most Commonly Used Techniques for Prompt Engineering

Now that we know what prompt engineering is, we can take a look at some of the best approaches to do it well. You'd be surprised how a simple change in phrasing can make the difference between a decent answer and a great one. Here, we'll cover some of the most common techniques to get ChatGPT to give you creative and useful responses you can use in a variety of situations.

Example 1: Teaching English to Non-native Speakers

Now, imagine you're a teacher trying to teach English to foreign students with various levels of competency in the language. In order to do so, you need to create a curriculum that accommodates your students and covers different topics and exercises that allow beginners to start right away and more experienced students to pick up where they left off. Let's look at two approaches we could take to build a curriculum for our English students. We're using Prompt 2-3 to get us started.

Prompt 2-3. Describing the problem

I'm an English teacher for foreign students and I'm looking for inspiration for creating a comprehensive curriculum.

Response 2-3 shows the response we got from ChatGPT for creating a curriculum for an English teacher.

Response 2-3. ChatGPT creates a curriculum for an English teacher

Creating a comprehensive curriculum for teaching English to foreign students involves several key steps. Here's a structured approach to help you design an effective and engaging curriculum:

Step 1: Define Objectives
Language Proficiency Levels: Identify the proficiency levels (beginner, intermediate, advanced) and align your curriculum with the Common European Framework of Reference for Languages (CEFR) or other standards like ACTFL.

19

Skills Focus: Determine the focus areas: speaking, listening, reading, writing, grammar, and vocabulary.
Cultural Competence: Include cultural knowledge and intercultural communication skills.

Step 2: Needs Analysis
Assess Student Needs: Conduct surveys or interviews to understand the specific needs, interests, and goals of your students.
Diagnostic Tests: Use placement tests to determine students' current proficiency levels.

Step 3: Develop Content and Materials
Thematic Units: Organize the curriculum into thematic units (e.g., daily life, travel, work, education).
Grammar and Vocabulary: Introduce grammar and vocabulary systematically and contextually.
Integrated Skills: Design activities that integrate multiple skills (e.g., a project that involves reading, writing, and speaking).

Step 4: Create Lesson Plans
Structured Lessons: Develop lesson plans that follow a consistent structure (warm-up, introduction, practice, production, review).
Diverse Activities: Include a variety of activities (e.g., discussions, role-plays, games, writing exercises).
Authentic Materials: Use authentic materials like news articles, videos, podcasts, and literature to enhance learning.

Step 5: Assessment and Evaluation
Formative Assessments: Use quizzes, homework, and in-class activities to monitor progress.

Summative Assessments: Design tests and projects to evaluate overall proficiency at the end of units or terms.
Self-Assessment: Encourage students to reflect on their learning and set personal goals.

Step 6: Incorporate Technology
Digital Tools: Utilize language learning apps, online dictionaries, and grammar checkers.
Interactive Platforms: Use platforms like Google Classroom, Edmodo, or Moodle to manage assignments and communication.
Multimedia Resources: Incorporate videos, interactive exercises, and audio recordings to enhance engagement.

Step 7: Continuous Improvement
Feedback: Regularly gather feedback from students and other teachers.
Professional Development: Stay updated with the latest teaching methods and resources through workshops, webinars, and courses.
Adapt and Update: Continuously revise and improve the curriculum based on feedback and new developments in language teaching.

Sample Curriculum Outline

Beginner Level
1. Unit 1: Introductions and Personal Information
Greetings, introductions, personal information questions.
Grammar: Simple present tense, pronouns.
Vocabulary: Family, numbers, days of the week.
Activities: Role-plays, flashcards, matching games.

2. Unit 2: Daily Routines
Describing daily routines and habits.
Grammar: Present simple tense, adverbs of frequency.

Vocabulary: Daily activities, time expressions.
Activities: Schedules, interviews, diary entries.

Intermediate Level
1. Unit 1: Travel and Transportation
Discussing travel plans, using public transportation.
Grammar: Future tense (will, going to), modal verbs.
Vocabulary: Travel-related terms, directions.
Activities: Trip planning, map reading, dialogues.

2. Unit 2: Health and Wellness
Talking about health issues and remedies.
Grammar: Present perfect tense, comparatives and superlatives.
Vocabulary: Body parts, symptoms, medical terms.
Activities: Doctor-patient role-plays, health surveys.

Advanced Level
1. Unit 1: Work and Career
Job interviews, discussing career goals.
Grammar: Conditionals, passive voice.
Vocabulary: Job-related terms, professional skills.
Activities: Mock interviews, resume writing, career planning.

2. Unit 2: Social Issues
   - *Debating social issues, discussing news articles.*
   - *Grammar: Reported speech, relative clauses.*
   - *Vocabulary: Social issues, opinion expressions.*
   - *Activities: Debates, article analysis, opinion essays.*

Resources and Tools
Textbooks and Workbooks: Choose reputable ESL textbooks that
align with your curriculum.
Online Resources: Websites like BBC Learning English, ESL
Library, and Duolingo.

Interactive Apps: Quizlet, Kahoot, and Anki for vocabulary practice.

Final Tips
Engage Students: Make lessons interactive and student-centered.
Real-life Practice: Encourage students to practice English outside the classroom.
Cultural Exchange: Incorporate cultural lessons and encourage sharing of cultural experiences.

By following these steps and continuously refining your approach, you can create a comprehensive and effective curriculum that meets the diverse needs of your students.

This is already a very helpful response. It provides multiple suggestions, an example lesson plan, and tips for constructing a curriculum successfully, even beyond the suggestions ChatGPT has given us here.

However, this single prompt won't create an entire lesson plan. See how we still need a teacher to construct something usable? The lesson plan ChatGPT provided us covers a broad range of topics; however, some of it may not be helpful for our complete beginners. For instance, beginners wouldn't be ready to discuss politics at any point in their first semester. Conversely, more advanced students only have two units to learn from, which couldn't last them an entire semester of learning.

Moreover, ChatGPT is essentially telling us to come up with the rest on our own. But we don't have to do it all ourselves. There's a way to get ChatGPT to take on the role of teacher and put more of the burden on the chatbot, which can help streamline our productivity and reduce the time spent brainstorming.

Make ChatGPT Know Its Role in the Conversation. Start with "You Are a..."

In this case, the trick to getting an even more detailed response from ChatGPT is to make it answer from the perspective of *being* the teacher, as shown in Prompt 2-4.

Prompt 2-4. Ensure ChatGPT knows its place!

You are an English teacher for foreign students that needs to create a curriculum that's comprehensive and easy to approach from various languages.

The difference between the two prompts is that your prompt makes ChatGPT play the role of the helpful assistant giving an English teacher some suggestions, which puts more burden on us for deciding how much to use and how much to disregard. However, in the second scenario, by telling ChatGPT to place itself in the shoes of the teacher itself, we're telling ChatGPT to give us an answer that comes from the viewpoint of a professional. ChatGPT is answering as someone that knows how to teach kids proficiently, as opposed to being the one giving professional ideas. Response 2-4 shows how ChatGPT answered.

Response 2-4. ChatGPT answers as a teacher creating a curriculum for teaching English

Creating a comprehensive and accessible English curriculum for foreign students involves structuring the course to cater to different proficiency levels, incorporating various teaching methods, and considering the linguistic and cultural backgrounds of the students. Here's a detailed curriculum outline:

Curriculum Overview
Levels: Beginner, Intermediate, Advanced

Components:
1. Vocabulary
2. Grammar
3. Reading Comprehension
4. Writing
5. Listening Comprehension
6. Speaking
7. Cultural Context

Beginner Level

1. *Vocabulary*
   - Basic words: Greetings, numbers, colors, days of the week, common objects
   - Thematic vocabulary: Family, school, food, body parts

2. *Grammar*
   - Simple sentence structure: Subject + Verb + Object
   - Present simple tense
   - Basic pronouns (I, you, he, she, it, we, they)
   - Articles (a, an, the)

3. Reading Comprehension
   - Short, simple texts: Dialogues, descriptions, daily routines
   - Comprehension questions focusing on identifying key information

4. *Writing*
   - Simple sentences and short paragraphs
   - Personal introductions, describing familiar objects or routines

5. *Listening Comprehension*
    - Listening to slow, clear speech: Simple conversations, instructions
    - Identifying basic information from spoken language

6. *Speaking*
    - Basic conversational phrases
    - Pronunciation practice
    - Role-playing simple scenarios (e.g., shopping, asking for directions)

7. *Cultural Context*
    - Introductions to English-speaking cultures
    - Common customs and polite expressions

Intermediate Level

1. *Vocabulary*
    - Expanding thematic vocabulary: Travel, hobbies, weather, jobs
    - Synonyms and antonyms

2. *Grammar*
    - Past simple and present continuous tenses
    - Comparative and superlative adjectives
    - Modals (can, could, may, might, should)
    - Introduction to complex sentences: Conjunctions (and, but, because)

3. *Reading Comprehension*
    - Short stories and articles
    - Answering comprehension questions and summarizing texts

4. *Writing*
   - Structured paragraphs and short essays
   - Descriptive writing, narrating events

5. *Listening Comprehension*
   - Listening to conversations and short stories
   - Answering questions about main ideas and details

6. *Speaking*
   - Extended conversations
   - Group discussions and presentations
   - Practicing intonation and stress in sentences

7. *Cultural Context*
   - Deeper understanding of English-speaking countries' traditions
   - Social norms and etiquette

Advanced Level

1. *Vocabulary*
   - Advanced vocabulary: Idioms, phrasal verbs, academic terms
   - Contextual usage and nuances

2. *Grammar*
   - Present perfect and past perfect tenses
   - Passive voice
   - Conditional sentences
   - Reported speech

3. *Reading Comprehension*
   - Longer, complex texts: Essays, news articles, literature excerpts
   - Critical analysis and interpretation

4. *Writing*
   - Essays, reports, and creative writing
   - Coherent and cohesive writing style
   - Argumentative and persuasive writing

5. *Listening Comprehension*
   - Understanding different accents and faster speech
   - Listening to lectures, debates, and news reports
   - Inferencing and identifying implied meanings

6. *Speaking*
   - Debates and discussions on various topics
   - Formal presentations
   - Fluency and accuracy in spoken English

7. *Cultural Context*
   - Cultural diversity in English-speaking countries
   - Analyzing cultural references in media and literature

Teaching Methods

1. *Interactive Activities:* Role-plays, group work, games
2. *Multimedia Resources:* Videos, audio recordings, online exercises
3. *Real-life Contexts:* Simulating real-world scenarios, field trips
4. *Feedback and Assessment:* Regular quizzes, feedback sessions, and progress reports

Additional Resources

   - *Textbooks and Workbooks:* Structured exercises and explanations

- *Online Platforms:* Duolingo, BBC Learning English, ESL websites
- *Apps:* Language learning apps for practice on the go
- *Cultural Exchange Programs:* Connecting with native speakers and cultural immersion activities

Curriculum Implementation

1. *Assessment:* Initial placement tests to determine proficiency levels.
2. *Customization:* Adapting the curriculum based on students' linguistic backgrounds and learning needs.
3. *Progress Tracking:* Regular assessments and feedback to monitor student progress.
4. *Support Systems:* Providing additional support through tutoring, language labs, and study groups.

By tailoring this curriculum to the specific needs of your students and incorporating various teaching methods, you can create an engaging and effective English learning experience for foreign students.

This answer is extremely expansive and goes more in depth. It covers grammar, suggests exercises for students, and creates a comprehensive plan for teaching students to read and write, speak, and listen at every stage of learning. This answer provides suggestions for using various tools for learning such as Duolingo and BBC English to complement the learning plan.

Both answers provided by ChatGPT are helpful, but making ChatGPT play the role of a teacher has given us a plan that's pretty close to being actionable. From here, ChatGPT can be prompted to create individual lessons based upon the suggestions it's already made, create templates for quizzes with answers, and more (however, be careful relying on the answers from ChatGPT for creating material that should be double-checked for accuracy; see **Notes** at the end of this chapter for more details).

Example 2: Summarizing a Paper on the History of DNA for a Fifth Grader and a College Student

Just about everyone struggles to find the right words. In a text message to a friend, you might end almost every sentence with "LOL," but that would be considered unprofessional in an email to your boss. Sometimes you know *what* you're trying to say, but not *how* you should say it. You're telling your boss that you had a pretty wild trip back home from vacation, but without putting "LOL" at the end, how do you keep the tone light?

Instead of having an existential crisis, we're going to turn to ChatGPT to transform our weak attempt at using the English language into something appropriate for different people in different situations.

In this case, we're going to look at how we can improve the language of a very plainly written summary. This will be the imaginary introduction to an imaginary paper on the discovery of DNA. Remember how we talked about ChatGPT being nondeterministic? Well, this is an example that really highlights that feature. We're going to take the same text and write it for two different audiences.

Set the Tone for a Fifth Grade Science Report

We already have a summary that we're asking ChatGPT to edit to make it more appropriate for different purposes. Prompt 2-5 is an example for a fifth grade science report.

Prompt 2-5. Getting ChatGPT to edit a science report summary

Rewrite this introductory summary to suit the tone of a 5th grader's science report:
###
This report explores the history of DNA discovery, starting with the early 20th-century advancements in genetics, highlighted by the rediscovery of Mendel's principles by Hugo de Vries, Carl Correns, and Erich von Tschermak. Thomas Hunt Morgan's work with fruit flies demonstrated that genes are located on chromosomes. Despite this, the chemical nature of genes remained unknown until mid-century. Key experiments by scientists like Oswald Avery, who identified DNA as the genetic material, and the Hershey-Chase experiments confirmed this. The 1950s were pivotal, with Maurice Wilkins, Rosalind Franklin, James Watson, and Francis Crick contributing to the discovery of DNA's double helix structure. Linus Pauling's earlier work on protein helices also played a role. These discoveries were foundational in understanding molecular biology and genetics, providing insights into the complex interactions of genes and proteins.

If you notice, we separated the context of the prompt (namely, the report summary) by three pound signs (###). This is a generally accepted technique that helps ChatGPT separate context and provide an answer that will meet your expectations, without getting anything confused.

We can see how ChatGPT answered this prompt in Response 2-5.

Response 2-5. ChatGPT helps us adjust the tone for a science report

```
My Science Report on the History of DNA Discovery
In this report, we're going to learn about how scientists
discovered DNA. It all started in the early 1900s when
three scientists, Hugo de Vries, Carl Correns, and Erich von
Tschermak, found out about Mendel's rules for how traits are
passed down in plants. Then, Thomas Hunt Morgan did experiments
with fruit flies and showed that genes are on chromosomes.

But even though they knew where genes were, nobody knew what
they were made of until much later. Scientists like Oswald
Avery did important experiments and found out that DNA is the
material that carries genetic information. Later, the Hershey-
Chase experiments confirmed Avery's findings.
The 1950s were super important for DNA discovery. Maurice
Wilkins and Rosalind Franklin used X-ray pictures to study DNA,
and James Watson and Francis Crick figured out that DNA has a
double helix shape, like a twisted ladder. Linus Pauling's work
on proteins also helped with understanding DNA's structure.
All these discoveries helped scientists understand molecular
biology and genetics better. They learned how genes and
proteins interact, which was a big deal for science.
```

This response expands on what we provided in the context of our prompt and does a good job of adjusting the tone to sound more juvenile. ChatGPT was made to specifically be able to understand and convey tone very well, and that shows here.

But there's even more we can do. Because ChatGPT is nondeterministic, with a small change to the prompt, we're going to get a completely different response in a completely different tone.

Set the Tone for an English Paper on the Discovery of DNA

Now we're going to use the same summary from Prompt 2-5, but we're going to alter it a bit so that we're asking ChatGPT to help us write for a completely different purpose. This time, we're looking for a more academic tone in the response (Prompt 2-6).

Prompt 2-6. Asking for an introductory paragraph for an English paper

Rewrite this in the tone of an introductory summary on the history of DNA and its discovery. Extra points are given out for making the tone of the writing engaging and easy to read, yet professional and academic.
###
This report explores the history of DNA discovery, starting with the early 20th-century advancements in genetics, highlighted by the rediscovery of Mendel's principles by Hugo de Vries, Carl Correns, and Erich von Tschermak. Thomas Hunt Morgan's work with fruit flies demonstrated that genes are located on chromosomes. Despite this, the chemical nature of genes remained unknown until mid-century. Key experiments by scientists like Oswald Avery, who identified DNA as the genetic material, and the Hershey-Chase experiments confirmed this. The 1950s were pivotal, with Maurice Wilkins, Rosalind Franklin, James Watson, and Francis Crick contributing to the discovery of DNA's double helix structure. Linus Pauling's earlier work on protein helices also played a role. These discoveries were foundational in understanding molecular biology and genetics, providing insights into the complex interactions of genes and proteins.

With our slight changes made, we can see just how well ChatGPT adheres to our request in Response 2-6.

Response 2-6. ChatGPT completely changes the tone of an English paper

The Journey of DNA Discovery: From Early Genetics to the Double Helix

The story of DNA discovery is a fascinating journey through the evolution of genetics and molecular biology. It begins in the early 20th century with the rediscovery of Gregor Mendel's principles of inheritance by scientists Hugo de Vries, Carl Correns, and Erich von Tschermak. These foundational ideas were crucial in establishing the field of genetics, but the quest to understand the physical basis of heredity had only just begun.

Thomas Hunt Morgan's groundbreaking work with fruit flies in the 1910s provided a key breakthrough. By demonstrating that genes are located on chromosomes, Morgan paved the way for a deeper exploration into the chemical nature of these hereditary units. Despite this, the specific molecule responsible for genetic information remained a mystery well into the mid-20th century.

A major milestone was achieved in the 1940s by Oswald Avery and his colleagues, who identified DNA as the substance that carried genetic information. This finding was further solidified by the Hershey-Chase experiments in 1952, which provided clear evidence that DNA, and not protein, was the genetic material in viruses.

The 1950s marked a pivotal era in DNA research. The collaborative efforts of scientists Maurice Wilkins, Rosalind Franklin, James Watson, and Francis Crick led to the discovery of the DNA double helix structure. This iconic model, revealed in 1953, elucidated how genetic information is stored and replicated. Franklin's critical X-ray diffraction images and Wilkins' insights were instrumental, alongside Watson and Crick's theoretical model building.

Additionally, Linus Pauling's earlier research on protein helices laid essential groundwork for understanding molecular structures, influencing Watson and Crick's approach to the DNA model.

These discoveries have profoundly impacted our understanding of molecular biology and genetics, providing invaluable insights into the complex interplay of genes and proteins. The journey from Mendelian genetics to the elucidation of the double helix represents one of the most significant scientific achievements shaping modern biology and medicine.

This response is wildly different from the first. It has a much more academic tone, just like we asked for. Yet it's not hard to get through or overly complicated. It completely delivers on everything we specified we needed for the tone of this summary. It's infinitely helpful to someone that struggles putting what they're thinking into words when the tone of what they've already written doesn't sound appropriate for the situation. It's hard to believe the two prompts we provided are only different by a few sentences!

Key Tips for Prompt Engineering

There are a few concepts that are essential to remember when working with ChatGPT, regardless of the purpose. While ChatGPT is an amazing tool that seems able to do anything, it's important to remember that it requires human supervision and management to create anything usable.

Therefore, let's look at the most common mistakes to be aware of when you're using ChatGPT.

Always Check Your Sources

There are a number of reasons that can cause ChatGPT to give you false information. For example, generally, you can expect that whenever you're using ChatGPT, it's been trained on information publicly available on the web a year ago from the current date. Therefore, if you're asking questions about sports scores for a particular game or recent election results in your prompts, chances are the information ChatGPT gives you won't be accurate.

ChatGPT is also limited to the sources of information it has available. If you're asking a question on a little-known topic, it's possible that ChatGPT isn't drawing from accurate sources. So, asking what would happen if you farted on Pluto might not result in a very scientifically accurate answer.

Finally, ChatGPT might simply be mistaken. Its algorithms draw from an incredibly large pool of data, and it's entirely possible that ChatGPT might mix up information when it provides an answer. If you're asking for a list of 100 best fantasy books, it might mix up and add sci-fi or historical fiction to the list. If you don't double-check, you might not otherwise notice the mistake.

This is why it's important to always do your own research when you're using ChatGPT when you're attempting to learn something from it or use what it's given you to present to other people. Otherwise, it's not ChatGPT that made the mistake, it's you for not double-checking the results.

ChatGPT Can Hallucinate, but What Does That Really Mean?

Wait. ChatGPT can *hallucinate*? Yes. It's actually very common.

Technically, hallucinating is simply when ChatGPT is making up facts that *it believes* are true. This can happen if you're in a long chat window that you've been reusing for a while. It might start pulling information from "sources" that don't exist. This issue can also occur if you ask ChatGPT a lot of questions about a subject it doesn't know about. You can experiment with this by asking it about a little-known book author and ask it details about them. Be really specific about the questions. It will probably start giving you facts about a person it's never heard of.

ChatGPT will almost certainly hallucinate if you *tell* it something is true, even if it's not. That's the difference between asking ChatGPT if a fact is true and telling it the fact *is* true. ChatGPT will rarely correct you unless you specifically ask for it, and even then, the response will likely favor what you already said.

Note   This issue can be greatly circumvented by using the recent features introduced in ChatGPT 4o that allows you to search the web for fact-checking. ChatGPT can also give you links so you can conduct your own research.

In a Prompt, ChatGPT Can Confuse the Difference Between Instructions and Information

Earlier, in Prompt 2-5, we gave ChatGPT instructions, then provided context (the necessary information) for the request. We separated the instructions from the information with three pound signs (###). This is because, otherwise, ChatGPT can get instructions confused for information and vice versa.

ChatGPT will also at times "forget" information you've given it earlier on in a conversation, which is another issue that can cause it to get mixed up. Because of this, it's wise to remind ChatGPT what you're talking about when the conversation has gone on for a while, even if you have to copy something you've already stated from earlier in the conversation and paste it back into your latest response.

Conclusion

In this chapter, you learned about prompt engineering and covered some of the most common conventions for getting helpful responses. We also became aware of some of the most common mistakes ChatGPT can make and how to avoid creating problems with the responses you get.

In the next chapter, we're going to learn more about using ChatGPT to create new content as a marketer, social media influencer, or just someone that needs ideas on what to make for dinner for the week.

CHAPTER 3

Prompting ChatGPT to Help You Create New Content and Get Ideas

Creating content can be both essential and tiring. Whether you're brainstorming ideas for a blog, a YouTube video, a business presentation, or even just planning meals for the week, it can be tough to stay creative and come up with fresh ideas. In our busy lives, where burnout is all too common, we could all use a little help getting inspired.

That's where ChatGPT comes in. This tool can assist with content creation by generating ideas when you're stuck. You can customize its responses to fit your specific needs, giving you a new perspective and helping to overcome creative blocks.

In this chapter, we will explore different situations where ChatGPT makes content creation easier. We'll also learn how to adjust our prompts to ensure ChatGPT provides unique and tailored responses to spark your creativity. Specifically, we'll cover the following:

© Lydia Evelyn 2025
L. Evelyn, *Making ChatGPT Work for You*, https://doi.org/10.1007/979-8-8688-1445-7_3

- Using ChatGPT to create a marking plan for a Twitch channel

- Asking ChatGPT to act as a PR manager to help a YouTuber prepare to get sponsorships

- Using ChatGPT to help with meal planning

Level Up a *Minecraft* Twitch Channel: Asking ChatGPT for a Marketing Plan

Twitch is a fantastic platform for the gaming community. It's an online service where people can watch and interact with live video broadcasts. Users, called streamers, broadcast themselves playing games or engaging in other activities, while viewers can watch, chat, and support their favorite streamers by following their channels, subscribing, or making donations. Gamers can find fans who love watching them play, build loyal followings, and even earn money through subscriptions and donations.

Turning your gaming hobby into a source of income is exciting, and Twitch is a great place for gamers to start. One of the most popular games on the platform is *Minecraft*. *Minecraft* is a popular video game where players can build and explore virtual worlds made up of blocks. It's like a digital sandbox where you can create anything you imagine, from simple houses to elaborate cities. Players can also go on adventures, fight monsters, and gather resources. *Minecraft* has both single-player and multiplayer modes, allowing people to play alone or with others on different servers, which are online worlds hosted by various people or communities.

While playing games like *Minecraft* on Twitch is fun and engaging, attracting and growing an audience is more challenging. Many content creators find that while they love gaming, they struggle with the marketing aspect needed to become successful influencers.

To address this challenge, we will explore how ChatGPT can help create a marketing plan for a Twitch streamer who mainly plays *Minecraft*. This plan will focus on gaining more followers, increasing channel monetization, and boosting income through donations and subscriptions.

Getting Relevant Marketing Ideas for a Twitch Channel

In Prompt 3-1, we're going to use ChatGPT to get useful advice on creating a marketing strategy.

Prompt 3-1. Give ChatGPT the scenario—you are a gamer on Twitch

```
You're a gamer on Twitch and stream Minecraft regularly.
You need to create a marketing plan to grow an audience and
improve engagement. Detail a marketing strategy tailored toward
nurturing a new channel aiming to draw the attention of the
Minecraft fanbase on Twitch.
```

This is a great prompt because, as we covered in Chapter 2, it's best to make ChatGPT play the role of the person needing to make a plan. For this reason, we told ChatGPT to answer from the perspective of a Twitch streamer. This prompt clearly conveys the goal of growing a Twitch audience from scratch and boosting engagement for *Minecraft* streams. Response 3-1 shows the results we got from ChatGPT using this prompt.

Response 3-1. A marketing plan from ChatGPT to help a *Minecraft* streamer on Twitch build their platform

Creating a successful marketing plan for your Twitch channel involves several key strategies, focusing on content quality, audience engagement, and promotion. Here's a detailed outline to help grow your audience and improve engagement for your Minecraft streams:

1. Branding and Identity

Channel Name and Logo: Ensure your channel name is memorable and reflective of your content. Create a professional logo and consistent branding elements (color scheme, fonts) for your Twitch channel and social media.
Channel Description: Write a compelling channel description that highlights what viewers can expect from your streams. Mention your streaming schedule, special events, and any unique aspects of your content.

2. Content Strategy

Regular Schedule: Establish a consistent streaming schedule. This helps viewers know when to tune in and builds routine viewership.
Stream Themes: Create specific themes or series for your Minecraft streams, such as "Survival Sundays," "Building Challenges," "Redstone Tutorials," or "Adventure Maps."
Interactive Content: Encourage viewer participation through polls, viewer challenges, or letting viewers influence in-game decisions
Collaborations: Collaborate with other Minecraft streamers to reach new audiences. Participate in joint streams, raids, and community events.

```
3. Engagement and Community Building

Chat Interaction: Actively engage with your chat. Respond to
comments, acknowledge new followers, and create a welcoming
environment.
Discord Server: Create a Discord server for your community. Use
it to share updates, host community events, and provide a space
for viewers to connect outside of streams.
Viewer Rewards: Implement a points or rewards system (using
Twitch's Channel Points) to incentivize viewer engagement.
Offer rewards like custom emotes, shout-outs, or in-game
actions.
```

This response is amazing for several reasons. It not only gives you advice that are on par with common trends within the streaming community on Twitch, like suggesting that you keep a streaming schedule, which is a well-known way to gain consistent viewers. It also suggests that you create themes for your streams so that viewers come back to continue the saga you've started and see the progress you made on *Minecraft*.

Even more impressive is that ChatGPT is using common vernacular that *Minecraft* players understand, like doing Surviving Challenges, Building Challenges, or Redstone Tutorials. Every *Minecraft* player knows how hard it is to play a *Minecraft* run without dying once. There are YouTube channels that show exclusively how to follow intricate blueprints for massive *Minecraft* mansions. And if you've played *Minecraft* long enough, you know no one even tries to do anything with Redstone without searching for a tutorial.

ChatGPT is also displaying a comprehension of Twitch's platform and how people connect, contribute, and expand their community via donations, Discord servers, and creating unique emotes for subscribers. Anyone that's used Twitch knows this is all pretty standard, but if you had no idea about any of these concepts beforehand, this is pretty darn helpful.

Let's take this even further and pry ChatGPT for more details.

Using ChatGPT to Stay Consistent with Creative Ideas That Engage Your Audience

In Prompt 3-2, we're going to use the same chat window with ChatGPT as before; therefore, ChatGPT will retain the context of the conversation we've had thus far, and we won't have to repeat ourselves.

Prompt 3-2. We're keeping it simple, silly

```
Come up with ideas for a themed stream series.
```

In this prompt, we're asking ChatGPT for help with one specific aspect of the advice it gave us earlier. We like the idea of creating a themed series of streams, so let's make ChatGPT expound on that to help us keep our audience engaged and, most of all, keep us from getting burned out from constantly coming up with ideas. Response 3-2 shows what we got back from ChatGPT.

Response 3-2. ChatGPT suggests how to create a series of themed streams

```
Creating a themed stream series can add excitement and draw
viewers who are interested in specific aspects of Minecraft
gameplay. Here are some ideas for a themed stream series:

1. "Minecraft Marvels: Build Challenges"
Theme: Creative Building
Description: Each stream focuses on building a marvel, such as
famous landmarks, fantasy castles, or futuristic cities. Invite
viewers to suggest themes or participate in build-offs.
Special Episodes: Collaborate with other streamers for joint
builds or competitions.
```

2. "Survival Saga: Hardcore Adventures"
Theme: Survival Mode
Description: Document your journey through Minecraft's survival mode on the hardest difficulty. Each episode presents new challenges, such as exploring dangerous biomes, defeating bosses, or building secure bases.
Interactive Elements: Allow viewers to suggest challenges or help make crucial survival decisions through polls.

3. *"Mystery Mansion: Roleplay Adventure"*
Theme: Roleplay and Storytelling
Description: Create a roleplay series set in a spooky mansion filled with puzzles, traps, and mysteries to solve. Develop a storyline with characters that viewers can interact with.
Special Features: Invite guest streamers to play characters or villains.

4. *"Minecraft Masterclass: Learn with Me"*
Theme: Tutorials and Education
Description: Each stream focuses on teaching viewers different Minecraft skills, such as redstone engineering, efficient farming, or advanced combat techniques.
Viewer Interaction: Conduct live Q&A sessions and invite viewers to submit their builds for critique and improvement tips.

5. *"Epic Expeditions: Exploring Minecraft Biomes"*
Theme: Exploration
Description: Set out on epic journeys to explore and document every biome in Minecraft. Each episode focuses on a different biome, highlighting its unique features, resources, and challenges.

Community Involvement: Organize group expeditions where viewers can join the adventure.

6. *"Villager Ventures: Trading and Economy"*
Theme: Trading and Villager Mechanics
Description: Dive deep into the mechanics of villager trading, build a thriving village, and optimize trading routes. Each episode can focus on different aspects, like building a marketplace or protecting villagers from raids.
Viewer Interaction: Create viewer-driven market days where viewers suggest trades and economic strategies.

7. *"Minecraft Mythbusters: Game Mechanics"*
Theme: Testing Game Mechanics
Description: Test and debunk popular Minecraft myths, glitches, and mechanics. Each stream can focus on different myths and involve experiments to see what's true.
Interactive Experiments: Encourage viewers to submit myths they want to see tested.

8. *"Pixel Art Parade: Artistic Creations"*
Theme: Pixel Art and Creative Mode
Description: Dedicate each stream to creating intricate pixel art designs or recreating famous artworks and characters in Minecraft.
Community Contributions: Showcase viewer-submitted pixel art and host contests for the best creations.

9. *"Redstone Revolution: Engineering Challenges"*
Theme: Redstone Engineering
Description: Focus on building complex redstone contraptions, from automated farms to intricate traps. Explain the mechanics and design process step-by-step.

Viewer Projects: Review and help troubleshoot viewer-submitted redstone projects.

10. "Minecraft Marathon: 24-Hour Survival"
Theme: Endurance and Charity
Description: Host a 24-hour marathon stream where you attempt to complete as many Minecraft challenges as possible, from defeating the Ender Dragon to completing advancements.
Charity Element: Partner with a charity and encourage donations throughout the stream, with milestones and goals linked to in-game challenges.

These themed stream series can keep your content fresh and engaging, attract new viewers interested in specific aspects of Minecraft, and build a loyal community around your channel.

These ideas are fantastic. As you can see, ChatGPT has given us ideas that are very relevant and aware of the Twitch and *Minecraft* gaming community. There are ideas for challenges that will keep the Twitch audience engaged and entertained, as well as tutorial ideas for people that watch because they're new to the game. ChatGPT has recommended content that experienced *Minecraft* players and newcomers alike would enjoy.

At this point, you can continue to toy around with ideas for even more content, or you can just save all of the content you got from ChatGPT and get started. As gamers would say, let's play.

So that we have ideas and a plan for our online gaming channel, let's change gears slightly and see what else ChatGPT can do for content creators.

Asking ChatGPT How You Can Glow Up and Approach Sponsors to Monetize a Beauty Vlogger's Viewership

Starting and maintaining a YouTube channel can be a fun hobby. Beauty vlogging is a popular venture that individuals begin simply because they enjoy showing off various creative ensembles, make up tutorials, or favorite products and brands. But for those that want to monetize their hobby, what was once fun becomes complex. One of the most reliable ways of monetizing their content is to make deals with sponsors, especially in the realm of beauty, where there are any number of products on the market with brands willing to make deals with beauty vloggers that stand out.

For any number of beauty bloggers, getting a good viewership and having a healthy relationship with potential brands for future sponsorship deals is the first step toward being able to monetize a beauty vlogging hobby. But first... how do you even get a sponsor in the first place? Well, let's ask ChatGPT.

Learning About Sponsorships on YouTube with ChatGPT

The prompt below is our first attempt to get ideas from ChatGPT on getting sponsors for a relatively new beauty vlogger on YouTube.

Prompt 3-3. A simple prompt for getting sponsors for a YouTube channel

You are a PR manager for a beauty vlogger just getting started on YouTube. How do you get started looking for sponsors?

Notice how we gave ChatGPT a role, yet again, in order to curate a response that will give us actionable advice, not just tips and suggestions that don't actually create a plan. Now let's look at the response, as is shown in Response 3-3.

Response 3-3. ChatGPT gives advice on securing brand deals for YouTube channels

Getting started with finding sponsors for a new beauty vlogger on YouTube involves several strategic steps. Here's a detailed plan to help you get started:

1. Build a Solid Foundation

a. Professional Branding:
- Create a professional media kit: Include the vlogger's bio, audience demographics, social media statistics, and engagement rates.
- Optimize social media profiles: Ensure all social media platforms are consistent and professional.
- Quality Content: Ensure the videos are high quality, engaging, and consistent in terms of style and posting schedule.

b. Audience Analysis:
- Know Your Audience: Collect data on your audience demographics, interests, and behavior. Tools like YouTube Analytics, Google Analytics, and social media insights are helpful.

2. Research Potential Sponsors

a. *Identify Relevant Brands:*
   - Niche Alignment: Look for brands that align with the vlogger's niche, values, and audience.
   - Competitive Analysis: Observe which brands are sponsoring similar vloggers.

b. *Create a Target List:*
   - Compile a list of potential brands and companies that are a good fit.

3. Craft a Compelling Pitch

a. *Personalized Outreach:*
   - Customize each pitch: Tailor the message to highlight how the vlogger's content aligns with the brand's goals.
   - **Highlight Audience Engagement**: Emphasize high engagement rates, audience loyalty, and any unique value the vlogger brings.

b. *Offer Value Propositions:*
   - Unique Selling Points: What sets the vlogger apart from others?
   - Collaboration Ideas: Propose creative and engaging content ideas for collaborations.

4. Utilize Networking and Platforms

a. *Leverage Existing Networks:*
   - Reach out to industry contacts: Use any existing relationships within the beauty and influencer industry.

b. *Use Sponsorship Platforms:*
   - Influencer Marketing Platforms: Sign up on platforms like FameBit, AspireIQ, or Grapevine Logic where brands look for influencers.
   - Marketplaces and Agencies: Consider joining influencer marketing agencies or marketplaces.

5. Consistent Follow-Up and Relationship Building

a. *Follow Up:*
   - Polite and Professional: Follow up on pitches respectfully and professionally if you don't receive a response initially.

b. *Build Long-Term Relationships:*
   - Deliver on Promises: Ensure that any sponsorship deals are executed flawlessly.
   - Feedback and Reporting: Provide brands with detailed reports on the performance of sponsored content.

6. Continual Improvement

a. *Gather Feedback:*
   - From Brands: Ask for feedback to improve future collaborations.
   - From Audience: Monitor audience reactions and engagement to sponsored content.

b. *Adapt Strategies:*
   - Stay Updated: Keep up with industry trends and adjust your strategies accordingly.
   - Experiment: Try different types of content and sponsorship models to see what works best.

Example Email Template for Pitching

Subject: Collaboration Opportunity with [Your Vlogger Name]

Hi [Brand Representative's Name],

I hope this email finds you well. My name is [Your Name], and I manage [Vlog Channel Name], a rising beauty vlogger on YouTube who is passionate about [specific niche, e.g., cruelty-free beauty, skincare routines, etc.].

With a rapidly growing audience of [number] subscribers and an engagement rate of [percentage], [Your Name's] content resonates with [brief description of the audience demographics]. We are impressed by [Brand's Name]'s commitment to [specific brand values or products] and believe there is a fantastic opportunity for a collaboration that could be mutually beneficial.

We would love to discuss how we can work together to create engaging and authentic content that highlights [Brand's Name] products to an enthusiastic and loyal audience. Attached is our media kit for your review.

Looking forward to the possibility of collaborating with you.

Best regards,
[Your Name]
[Your Contact Information]

By following these steps and continually refining your approach, you can effectively start securing sponsorships for the new beauty vlogger.

Well, this response not only gave us an explanation on how to approach sponsors but also how to do the appropriate research, find the right contacts, and initiate a conversation in a professional way.

As previously mentioned, this advice is very actionable. You can save this conversation and use it as a guide right away. But let's think ahead for a second. What if we get a response? What can we expect? Let's get a little more information before we dive right in.

Preparing to Maintain Professional Relationships with Advice from ChatGPT

In Prompt 3-4, we're asking for the next steps in a hypothetical where we obtain a brand deal from acting upon the advice from the previous step. This is helpful for planning ahead and having a strategy already in place for conducting business for your channel so you can maintain a professional relationship with your brands.

Prompt 3-4. Make a plan for successful correspondences

What can I, as a beauty vlogger, expect in a positive response from a brand I've contacted? How do I negotiate a proper deal?

So, let's look at what ChatGPT has to say, as you'll see in Response 3-4.

Response 3-4. Getting next step ideas for obtaining a business deal

As a beauty vlogger, receiving a positive response from a brand you've contacted is an exciting step. Here's what you can expect in a positive response and how to negotiate a proper deal:

53

Positive Response Indicators

1. *Interest in Collaboration:*
   - The brand expresses enthusiasm about your content and audience.
   - They mention specific aspects of your channel that align with their brand values and products.

2. *Proposal for Next Steps:*
   - They might suggest a meeting or a call to discuss potential collaboration ideas in detail.
   - They could outline initial thoughts on how they envision the partnership.

3. *Request for Additional Information:*
   - They may ask for more detailed Analytics about your audience.
   - They might request to see examples of past sponsored content or your media kit.

Negotiating a Proper Deal

1. *Understand Your Worth:*

   - Know Your Metrics: Be clear about your reach (subscribers, followers), engagement rates (likes, comments, shares), and viewer demographics.
   - Industry Rates: Research standard rates for influencers at your level. Tools like Social Bluebook or Influencer Marketing Hub can help.

2. *Define the Scope of Work:*

   - Content Type: Specify the type of content you will create (e.g., dedicated video, mention, Instagram post, story).

- Deliverables: Outline the exact deliverables, including the number of posts, platforms, and any other requirements (hashtags, mentions, links).

3. *Compensation:*

- Monetary Payment: Clearly state your rates for the proposed work.
- Product Compensation: If the brand offers products, consider if they complement monetary compensation or if they suffice for smaller collaborations.
- Hybrid Deals: Sometimes a mix of monetary payment and products can be negotiated.

4. *Usage Rights and Exclusivity:*

- Content Usage: Specify if the brand can use the content on their own channels and for how long.
- Exclusivity: Discuss if there will be any exclusivity clause preventing you from working with competitor brands for a certain period.

5. *Timeline and Deadlines:*

- Content Schedule: Agree on a timeline for content creation and posting.
- Approval Process: Clarify how the approval process will work and how many rounds of revisions are included.

6. *Performance Metrics and Reporting:*

- Tracking Performance: Determine how the performance of the sponsored content will be tracked and reported back to the brand.
- Delivering Results: Agree on a method for delivering these results, such as screenshots of Analytics or access to performance data.

Example Response to a Positive Brand Inquiry

Hi [Brand Representative's Name],

Thank you for your positive response and the opportunity to potentially collaborate with [Brand's Name]. I'm excited about the possibility of working together and believe our audiences would be highly interested in [Brand's Name] products.

To better understand how we can create the most effective and engaging content, could we schedule a call to discuss the details? I'd love to share some ideas on how we can highlight [Brand's Name] in a way that resonates with my audience.

In preparation for our discussion, here are some initial thoughts on the collaboration:

- Type of Content: [e.g., YouTube tutorial featuring [Product], Instagram stories, etc.]
- Deliverables: [e.g., One dedicated YouTube video, three Instagram posts, etc.]
- Compensation: [e.g., My rate for this scope of work is $X, and I am open to discussing product compensation as well.]
- Timeline: [e.g., Proposed timeline for content creation and posting.]
- Additional Information: [e.g., I have attached my media kit for your review.]

Looking forward to hearing from you and discussing this exciting opportunity further.

Best regards,
[Your Name]
[Your Contact Information]

During Negotiation

1. Be Clear and Professional: Maintain professionalism in all communications. Be clear and concise about your needs and expectations.
2. Be Open to Compromise: Be willing to negotiate terms that work for both parties. Flexibility can often lead to a more fruitful collaboration.
3. Get Everything in Writing: Ensure all agreed terms are documented in a contract to protect both parties and outline responsibilities clearly.

By following these steps, you can effectively negotiate a fair and beneficial deal with brands interested in collaborating with you.

This is excellent advice and a very helpful conversation overall, especially coming from having known nothing at all about making brand deals or seeking out sponsorships. Still, ChatGPT can make mistakes, and it's important to conduct your own research, as ChatGPT can never replace professional advice.

However, ChatGPT can make your research tasks easier. With models like ChatGPT 4o, you can now ask ChatGPT to search the web for results, and it will cite its sources in its response. This allows you to find out for yourself where ChatGPT is getting information and double-check that you're reading from reputable sources. Let's see this in action by asking ChatGPT what we should know about making deals with trustworthy brands.

Asking ChatGPT to Cite Sources and Help Us Find Reputable Brands for Sponsorships

Prompt 3-5 is short, sweet, and to the point. We're asking ChatGPT to search for the information and give us links that allows us to check its work, so to speak, so we can evaluate the information for ourselves straight from the source. ChatGPT's answer is in Response 3-5.

Prompt 3-5. Getting ChatGPT to cite sources for accurate responses

What beauty brands are known for being trustworthy and fair with influencers? Cite your sources in your response.

Response 3-5. ChatGPT provides up-to-date information by searching the web

As a beauty vlogger looking to partner with trustworthy and fair brands, there are several companies known for their positive and supportive relationships with influencers. Here are some of the top brands you can consider:

1. Sephora: Sephora is renowned for its comprehensive influencer programs, particularly the Sephora Squad. This initiative offers influencers the chance to collaborate with a wide array of brands and participate in exclusive events. Sephora is known for its professionalism and fair compensation practices.

2. Fenty Beauty: Founded by Rihanna, Fenty Beauty has revolutionized the beauty industry with its emphasis on inclusivity and diversity. The brand is highly praised for its fair treatment of influencers, providing ample opportunities for collaboration and ensuring that influencer content aligns with their message of empowerment and authenticity.

3. Rare Beauty: Created by Selena Gomez, Rare Beauty focuses on promoting self-expression and mental health. The brand is known for its supportive and fair collaborations with influencers, emphasizing authenticity and inclusivity in their partnerships.

4. L'Oréal: As one of the largest beauty companies globally, L'Oréal collaborates with a diverse range of influencers across its many brands, including Maybelline and NYX Professional Makeup. They are known for their structured and professional approach to influencer marketing, ensuring fair compensation and clear expectations.

5. MAC Cosmetics: MAC is a favorite among professional makeup artists and beauty enthusiasts. They have a long history of working with influencers and are known for their creative collaborations and fair treatment of partners.

6. La Roche-Posay: This skincare brand is respected for its dermatologist-recommended products and ethical approach to influencer marketing. They frequently collaborate with influencers to promote products designed for sensitive skin and other specific skin concerns.

7. The Ordinary: Part of Deciem, The Ordinary is highly regarded for its transparency and ethical marketing practices. They offer straightforward, science-backed skincare products and are known for fair and honest collaborations with influencers.

8. CeraVe: Known for its dermatologist-developed skincare products, CeraVe has a strong reputation for treating influencers fairly. They focus on educational content and often collaborate with skincare professionals and enthusiasts.

9. ColourPop: ColourPop is celebrated for its vibrant and affordable products. The brand frequently collaborates with influencers and celebrities, providing ample opportunities for creative and engaging partnerships.

When negotiating deals with these brands, ensure that you discuss key elements such as compensation, deliverables, usage rights, and timelines to establish a clear and fair agreement. Document all terms in a contract to protect both parties and maintain a professional relationship.

Sources:
https://www.gcimagazine.com/consumers-markets/
news/22885121/20232024-top-beauty-influencers-brands
https://hypeauditor.com/blog/top-15-beauty-brands-that-work-
with-influencers/

Here, ChatGPT has provided us with links we can check ourselves to see where the information came from, and from there, we can check the sources of *those* links to ensure the information we're working with is reputable.

But enough about business. Sometimes our needs are a lot simpler than trying to create a steady side hustle. Some of us have enough on our plates, literally, just trying to come up with ideas for dinner.

Plan Meals and Stop Stressing About Dinner Ideas with ChatGPT

Everyone is busy nowadays. Every day we have bills to pay, elections to worry about, a planet to save, and all that's before we've even had dinner yet. Coming up with dinner ideas is hard when you're always thinking about a thousand other things.

It's even harder when you have dietary needs. Some people can't eat gluten; other people can't have sugar. Vegans can't eat meat, and pescatarians need to eat, too. Maybe you're trying to lose weight and need lower calorie meals, or you're trying to keep up your protein and need help making meal plans that cover all of your macros.

Instead of adding basic meals to your list of problems, use ChatGPT so that making dinner isn't the cause of yet another existential crisis.

Preparing a Meal Plan for the Week Ahead When You Have Dietary Restrictions

Let's challenge ChatGPT with a hard one (Prompt 3-6). Meet the Johnsons, a family of impossibly picky eaters:

Prompt 3-6. Giving ChatGPT the impossible task of pleasing picky eaters

Create a meal plan for the Johnsons, a family of five. However, Paula is lactose intolerant, Michael is vegan, and Liam is on a keto diet. Anna and John don't care either way, but none of the family members like chicken. Plan a week of dinners that accommodates the whole family.

No wonder the Johnsons need help with dinner ideas. It's hard to imagine how they pick out restaurants! Let's see if ChatGPT is up for the challenge, as you can see in Response 3-6.

Response 3-6. A meal plan for the Johnsons, against all odds

Creating a meal plan for the Johnson's family requires some careful planning to accommodate Paula's lactose intolerance, Michael's vegan diet, and Liam's keto diet, while ensuring Anna and John have meals they enjoy too. Here's a week of dinners that fit these requirements:

Monday
Stuffed Bell Peppers
- Paula & Liam: Bell peppers stuffed with cauliflower rice, ground beef, and vegetables, topped with dairy-free cheese.
- Michael: Bell peppers stuffed with quinoa, black beans, corn, tomatoes, and avocado.

Tuesday
Zucchini Noodles with Pesto
- Paula & Liam: Zucchini noodles with avocado pesto, cherry tomatoes, and grilled shrimp.
- Michael: Zucchini noodles with avocado pesto, cherry tomatoes, and sautéed tofu.

Wednesday
Baked Tofu & Vegetable Stir-Fry
- Paula & Liam: Stir-fry with tofu, broccoli, bell peppers, snap peas, and a soy-ginger sauce, served over cauliflower rice.
- Michael: Same stir-fry with tofu.

Thursday
Cauliflower Crust Pizza
- Paula & Liam: Cauliflower crust pizza with tomato sauce, dairy-free cheese, pepperoni, and vegetables.
- Michael: Cauliflower crust pizza with tomato sauce, vegan cheese, and a variety of vegetables.

Friday
Lentil and Vegetable Stew
- Paula & Liam: Stew made with lentils, carrots, celery, tomatoes, and kale, seasoned with herbs and spices.
- Michael: Same lentil and vegetable stew.

Saturday

Eggplant Lasagna
- Paula & Liam: Eggplant slices layered with ground beef, dairy-free ricotta, and marinara sauce.
- Michael: Eggplant slices layered with spinach, mushrooms, vegan ricotta, and marinara sauce.

Sunday

Tacos
- Paula & Liam: Lettuce wrap tacos with ground beef, avocado, salsa, and dairy-free cheese.
- Michael: Lettuce wrap tacos with black beans, corn, avocado, salsa, and vegan cheese.

Notes

- For Anna and John, each meal can be adapted easily by adding regular cheese or other toppings they enjoy.
- Ensure that all sauces, especially pesto and marinara, are dairy-free for Paula and vegan for Michael.
- Liam's keto diet can be managed by ensuring his portions are high in healthy fats and low in carbs, such as using cauliflower rice or zucchini noodles instead of grains.

This meal plan should provide a balanced and enjoyable week of dinners for the entire Johnson's family, accommodating all their dietary needs and preferences.

As you can see, ChatGPT has accommodated every family member in its response. Let's take a look at what the grocery list for the week would look like. After all, who wants to make multiple trips to the grocery store when we can get everything at once and go back to watching anime on Netflix?

Making a Quick Grocery List so You Have Everything on Hand

Obviously, ChatGPT knows what should go in the meals it's suggested, so let's avoid going through the trouble of coming up with the grocery list. Below is a short prompt in the same conversation that will allow us to create a grocery list based upon the meals we've decided on.

Prompt 3-7. Make an easy grocery list

Create a grocery list for the week based upon the meals you've suggested.

Response 3-7 is the answer we got back from ChatGPT.

Response 3-7. A grocery list you didn't have to write yourself

Sure, here's the grocery list for the week based on the suggested meals:

Produce
- Bell peppers (8)
- Cauliflower heads (3)
- Zucchini (8)
- Avocados (6)
- Cherry tomatoes (3 pints)
- Broccoli (2 heads)
- Snap peas (1 bag)
- Carrots (6)
- Celery (1 bunch)
- Kale (1 bunch)
- Eggplant (4)
- Lettuce (2 heads or 1 bag of large leaves for wraps)
- Mushrooms (2 cups)

- Spinach (1 bag)
- Garlic (1 bulb)
- Ginger (1 piece)
- Fresh herbs (basil, parsley, cilantro)
- Onions (6)
- Limes (4)

Proteins
- Ground beef (3 pounds)
- Tofu (3 blocks)
- Shrimp (2 pounds)
- Black beans (4 cans)
- Lentils (2 bags)
- Pepperoni (1 package)
- Vegan ricotta (1 container)
- Dairy-free cheese (3 packages)
- Vegan cheese (3 packages)

Grains and Legumes
- Quinoa (1 bag)
- Cauliflower rice (frozen or fresh, 3 bags)
- Corn (frozen or canned, 2 cans)

Sauces and Condiments
- Tomato sauce (2 jars)
- Marinara sauce (2 jars)
- Soy sauce (1 bottle)
- Ginger sauce (1 bottle)
- Salsa (1 jar)
- Vegan pesto (or ingredients to make: basil, pine nuts, olive oil)

Spices and Seasonings
- Salt
- Pepper
- Italian seasoning
- Cumin
- Chili powder
- Garlic powder
- Onion powder

Baking and Cooking Essentials
- Olive oil
- Coconut oil
- Vinegar (apple cider or balsamic)
- Nutritional yeast

Miscellaneous
- Cauliflower pizza crusts (frozen, 4)
- Almond milk or other dairy-free milk (for cooking)

This list should cover all the ingredients needed for the dinners throughout the week. Be sure to check your pantry for any items you might already have to avoid duplicates.

Obviously, the Johnsons might have a couple of these ingredients on hand already, so they can take out anything they already have, as ChatGPT suggests. However, this is an incredibly handy tool that takes the mental effort off of anyone stressed about making plans for the week.

But what happens when you're caught off guard? Sometimes you forget to make plans until it's already too late, your cabinets are empty, and you want something quick. Let's see if ChatGPT can help out.

Asking ChatGPT for a Dinner Idea When You're Low on Ingredients

We've all been there before. It's 8pm, it's a holiday, everything's closed, the fridge is empty, and you're hungry. You realize there are a few corner stores open, but you have no idea what to grab that would come together into a decent meal. You can do what you normally do and try to piece together a meal from random ingredients. But if you're just tired of eating the same three recipes on rotation and want to try something new, then toss the problem to ChatGPT. In Prompt 3-8, we ask ChatGPT to help us with our problem.

Prompt 3-8. Laying out the issue for ChatGPT

```
Most of the stores are closed except for some corner stores
with basic ingredients. I'm too tired to make anything too
complicated for dinner. I'm willing to pick up or order a few
ingredients from Uber.
```

What we've done is lay out the issue in simple terms and expressed what we are and aren't willing to do. For example, we don't want to spend too much time in the kitchen. Let's see if ChatGPT can come to the rescue (Response 3-8).

Response 3-8. Super simple meals with minimum ingredients

```
Here are a few easy dinner ideas using basic ingredients you
might find in corner stores:
```

1. Quick Pasta

Ingredients:
- Pasta
- Canned tomatoes or tomato sauce
- Garlic
- Olive oil
- Salt and pepper

Instructions:
1. Cook the pasta according to package instructions.
2. While the pasta is cooking, heat some olive oil in a pan and sauté minced garlic until fragrant.
3. Add the canned tomatoes or tomato sauce to the pan and simmer for 10 minutes. Season with salt and pepper.
4. Toss the cooked pasta in the sauce and serve.

3. Egg Fried Rice

Ingredients:
- Cooked rice
- Eggs
- Soy sauce
- Frozen or canned vegetables (peas, carrots, corn, etc.)
- Oil

Instructions:
1. Heat some oil in a pan or wok.
2. Scramble the eggs and set them aside.
3. Add more oil to the pan and stir-fry the vegetables until tender.
4. Add the cooked rice and stir-fry for a few minutes.

5. Mix in the scrambled eggs and soy sauce. Stir well
   and serve.

4. Tuna Salad Sandwich

Ingredients:
- Canned tuna
- Mayonnaise
- Bread
- Lettuce (optional)
- Salt and pepper

Instructions:
1. Drain the canned tuna and mix it with mayonnaise. Season
   with salt and pepper.
2. Spread the tuna mixture on a slice of bread. Add lettuce if
   you have it.
3. Top with another slice of bread and serve.

5. Veggie Stir-Fry

Ingredients:
- Mixed vegetables (fresh, frozen, or canned)
- Soy sauce
- Garlic
- Oil
- Rice or noodles (optional)

Instructions:
1. Heat oil in a pan or wok.
2. Sauté minced garlic until fragrant.
3. Add the mixed vegetables and stir-fry until cooked through.
4. Season with soy sauce.
5. Serve as is or over rice/noodles.

These recipes are simple, quick, and require minimal
ingredients and effort. Enjoy your meal!

As you can see, ChatGPT managed to provide some very simple options for low-effort dishes that seem easy to accomplish and don't require too many ingredients. On an already tough day, this can be a lifesaver. A simple prompt like this could definitely help someone end up with a more complete meal for dinner than, say, a jelly sandwich and some crackers.

Conclusion

In this chapter, we've looked at how we can use ChatGPT to streamline content creation, from coming up with marketing plans for a Twitch streamer to creating a strategy to contact sponsors to planning for dinner for truly any need.

In the next chapter, we're going to look at how we can teach ChatGPT with large amounts of data, or context, to condense information into something easier to understand and overall streamline productivity at work.

CHAPTER 4

Teaching ChatGPT to Create Content in Different Styles

With more and more people incorporating ChatGPT into their workflow in the office, many get frustrated with responses that sound inauthentic. This is especially true when we're talking about trying to recreate a writing style—something that's distinct among writers of all kinds. A writing style is difficult to replicate, and a chatbot's attempt typically sounds like a cheap imitation. Without a doubt, if you just ask ChatGPT to write about any subject without any other context, the answer you get will be very dry and uninspired.

The truth is, ChatGPT can take on a more natural-sounding writing style if it's *taught* to do so. In this chapter, we're going to learn how to teach ChatGPT to write in any style for writing social media content, writing emails, help with time management, and help write blog posts. We're also going to be looking at how to use the web search feature in ChatGPT to provide more background and context to its responses.

© Lydia Evelyn 2025
L. Evelyn, *Making ChatGPT Work for You*, https://doi.org/10.1007/979-8-8688-1445-7_4

Here's what we're going to cover in this chapter:

- Teaching ChatGPT to be your social media manager, so that you can create eye-catching social media posts and keep up with trends to stay relevant as a small business owner

- Instructing ChatGPT on how to write emails in the voice of someone else, for example, for an administrative assistant writing emails for a boss that doesn't have time to write them themselves

- Using ChatGPT for time management by training it on your schedule

- Asking ChatGPT to use web search to find information for a blog post and training it to write in the style of the blog's author

Teaching ChatGPT to Be Your Social Media Manager

In this day and age, it's critical to understand how to engage an audience on social media, as well as keep up with trends that the algorithms favor based on the success of similar advertising campaigns in your field. Although this is particularly a problem for social media managers, this is especially crucial for small business owners that may depend on having a social media following engaged for sales and business interest.

When used correctly, ChatGPT can be used to make social media management easier and even more efficient than if you did it on your own. Not only can you provide ChatGPT with content from other social media influencers with similar products or business models as your own, but you can also include **images** in your prompt. This allows you to better understand, for example, how certain shots and compositional elements in product photos perform better than others. Not only that, but you can ask ChatGPT for advice on how to display your own products to solicit a positive response from your followers and encourage more engagement.

Creating a Better Following on Instagram

For restaurant owners, a strong online presence makes it easier for customers to find them. Having a presence on Google Maps makes it easy for customers to find the restaurant before they leave the house. On the website, customers can read the menu before they've even stepped into the establishment. And with Instagram, customers can see how good the food looks and arrive already excited to eat.

Taking a picture of good food may sound simple, but there's an art to it. We're going to have ChatGPT explain to us how that works by giving it an example of a post that found great success and asking how we can reproduce the results.

For reference, we're uploading the image shown in Figure 4-1 to ChatGPT in our next prompt.

Figure 4-1. *A beautiful photo of a restaurant breakfast mealPhoto credit: RitaE on Pixabay*

You can send photos to ChatGPT that can be analyzed and used as reference for a number of purposes. In this case, we're uploading an image like one you might find on Instagram. Imagining this is a successful Instagram from a restaurant promoting items on their menu, we're going to ask ChatGPT to help us garner similar results for our own posts.

Prompt 4-1. Asking ChatGPT for advice on making a post for a restaurant on Instagram by giving it a viral post as an example

```
I own a restaurant and I'm trying to grow an audience on
Instagram to bring more attention to my business. I'm seeing
posts that are getting attention in the same field, and I'm
trying to figure out how to make these strategies work for
me. What equipment do I need to take photos like the one I've
```

uploaded? What should I post? How should I prepare the photo? What should I use in the caption? Give me some advice.

This is the caption for the photo I've just uploaded:
###
Amidst the quiet hush of a morning well-spent, delicate crepes arrive like soft whispers—lightly dusted with powdered sugar, scattered with plump blueberries, and served alongside the golden glow of apricot jam. A gentle beginning to the day, as sweet as it is serene.

As you can see, we've separated the information from our request with "###" and included the caption so that ChatGPT can give us detailed advice. Let's look at Response 4-1.

Response 4-1. ChatGPT gives us very useful advice so we can learn how to make successful posts on Instagram

To achieve a polished, engaging Instagram presence for your restaurant, you'll need to focus on high-quality photos, thoughtful presentation, and compelling captions. Here's a breakdown of strategies, equipment, and best practices to elevate your posts:

Equipment Needed:

1. Camera: While modern smartphones (like iPhones or high-end Android phones) work well, consider a DSLR or mirrorless camera (like the Canon EOS M50 or Sony Alpha a6400) for even more professional shots.
2. Lighting: Good lighting makes a massive difference. Invest in a ring light or a softbox setup to achieve a natural and balanced light. Try shooting in soft natural light if available.

3. Tripod: Ensures stability and helps avoid blur, especially useful for shots that require precision or a consistent angle.

4. Backdrop: Consider having clean, stylish backdrops (wooden boards, stone slabs, or textured fabric) to create a more attractive setting for your dishes.

5. Editing Software: Tools like Adobe Lightroom, VSCO, or Snapseed are excellent for editing photos to adjust lighting, sharpness, saturation, etc., enhancing the final look.

Tips for Preparing and Taking Photos:

1. Presentation Matters: Ensure your plating is artistic and appetizing, focusing on colors and textures. Highlight the star of the dish clearly.

2. Composition: Consider the rule of thirds, negative space, and angles. Overhead shots, close-ups, and angled side shots often work well for food.

3. Props and Context: Surround your dish with relevant props (e.g., wine glasses, cutlery, spices). Be careful not to clutter the shot.

4. Lighting: Use soft, diffused light to reduce harsh shadows. Side or natural light often enhances food's appearance.

5. Details: Show texture (crispy edges, glistening sauce). Use a macro lens if focusing on intricate details.

Caption Strategies:

1. Tell a Story: Connect your dish to a season, cultural experience, or history (as you did in your uploaded caption).

2. Descriptive Words: Use sensory language (e.g., "crisp," "rich," "aromatic") that lets readers almost taste the food.

3. Call-to-Action (CTA): Encourage engagement, such as "Have you tried this?" or "Tag someone who would love this!"

4. Hashtags: Use relevant ones (e.g., #GameSeasonEats, #FoodieFavorites, #SeasonalSpecials) to extend reach.

Caption Improvement for Your Uploaded Photo:

Your caption already creates a vivid atmosphere. Here's a suggestion to enhance engagement further:

• Add a Question: "What's your favorite accompaniment to game dishes?"

• Shorten/Refine for Impact: "Game season brings bold flavors! Our plump partridge with warm, clove-infused bread sauce and aromatic quince jelly is a comforting classic. Would you pair it with a crisp wine or ale? Let us know!"

Post Ideas:

1. Behind-the-Scenes: Show prep, plating, or sourcing fresh ingredients.

2. Chef Insights: Have your chef discuss cooking techniques or unique ingredient stories.

3. Seasonal Specials: Showcase new or limited-time dishes.

4. Customer Spotlights: Share customer photos and testimonials.

5. Interactive Content: Polls, quizzes, and story takeovers to encourage interaction.

Combining these strategies should help grow and engage your Instagram audience, drawing attention to your restaurant.

Using ChatGPT for Feedback to Critique an Instagram Post

ChatGPT has given us some really good advice. In fact, let's try to implement it! In Figure 4-2, you'll see that I've included what I believe to be an extremely professional post of an incredibly appetizing dish that I believe will easily earn thousands of likes and follows on Instagram.

Figure 4-2. *A delicious-looking post that will surely go viral on Instagram*

Likewise, I've also crafted what I believe to be a very relatable and engaging caption to go with the photo in Listing 4-1.

Listing 4-1. A caption to match a soon-to-be viral Instagram post

```
Um, here's an egg. We have scrambled eggs, but I just took a
picture of a sunny-side up one, I don't know. Anyway, we have
some other dishes at the restaurant. Come check them out if you
have time. #idunno #please #eat
```

Now, I'm pretty sure this post will be a hit, but just in case, lets ask ChatGPT to compare our post to the viral post we're comparing ourselves to and see where there may be room for improvement. Please note that this prompt includes the image shown in Figure 4-2. That said, let's check out Prompt 4-2.

Prompt 4-2. Asking ChatGPT for feedback on a potential Instagram post for a restaurant

```
Alright, I have a photo of a dish from my restaurant. Here is
my idea for the caption:
```

```
###
```

```
Um, here's an egg. We have scrambled eggs, but I just took a
picture of a sunny-side up one, I don't know. Anyway, we have
some other dishes at the restaurant. Come check them out if you
have time. #idunno #please #eat
```

```
###
```

```
Give me some feedback and let me know how well I've implemented
the points we discussed to help me grow a following on
Instagram
```

So, what did ChatGPT think? Take a look at Response 4-2.

Response 4-2. ChatGPT tells us how to improve our Instagram post

Your caption needs some adjustments to better align with the strategies we discussed for engaging your audience and promoting your restaurant effectively on Instagram. Here are some suggestions:

Caption Feedback and Revision

Original Caption:
Um, here's an egg. We have scrambled eggs, but I just took a picture of a sunny-side up one, I don't know. Anyway, we have some other dishes at the restaurant. Come check them out if you have time. #idunno #please #eat

Revised Caption:
"Start your day sunny-side up with our delicious egg dishes! 🔍 ❄️ Whether you prefer them scrambled, poached, or sunny-side up, we've got you covered. Come visit us and explore our full breakfast menu! ⚙️ 🍽️ #foodie #egglover #gastronomy #NameOfTheRestaurant"

Key Improvements:
1. Engagement: The revised caption encourages followers to start their day with your egg dishes, creating a positive and inviting tone.
2. Clarity and Confidence: It clearly states what the post is about and does so confidently.
3. Emojis: Appropriate use of emojis adds a friendly and engaging touch.
4. Hashtags: Relevant hashtags help reach a broader audience and attract egg lovers and foodies.
5. Call to Action: Encourages followers to visit the restaurant and try the full breakfast menu.

Additional Tips for the Photo:
- Lighting: Ensure the dish is well-lit. Natural light from a window can enhance the appearance.
- Plating: Consider adding garnishes like fresh herbs or a sprinkle of spices to make the dish more visually appealing.
- Angle: Try different angles to find the most flattering perspective for the dish. Overhead shots often work well for food photography.

Next Steps:
1. Edit the Photo: Use a photo editing app to adjust brightness, contrast, and saturation to make the colors pop.
2. Repost with Revised Caption: Update the caption with the revised version to better attract and engage your audience.

By implementing these strategies, you can create more appealing and engaging Instagram posts to grow your restaurant's following and attract more customers.

This is great! Not only can we get advice from ChatGPT on how to make a good Instagram post, but we can also get feedback on our attempts to improve. That said, I'm not so sure my restaurant would survive on dishes like mine, and I should stick to writing books.

Teaching ChatGPT to Help an Administrative Assistant

In the workforce, an administrative assistant is often required to send emails on behalf of a manager that doesn't have time to do so themselves. Let's look at a scenario where an administrative assistant might need to send a company-wide message on behalf of their manager.

For example, a tech company manager might need to communicate to the rest of the company the launch of a new security measure requirement, like adding two-factor authentication to their emails, which is a common procedure nowadays. If you didn't know, Outlook actually has a feature that makes this easier. It's called the **Send as** feature, which can be set up by an IT manager for any company that utilizes Microsoft for their email systems.

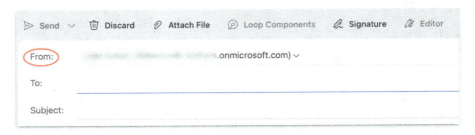

Figure 4-3. *An example of what it looks like when someone is given permission to send an email as someone else*

What Happens When You Ask ChatGPT to Write the Email Without Any Additional Information?

As stated before, a lot of people get frustrated when ChatGPT responds with content that sounds stiff or robotic. Let's take a look at how that happens by asking ChatGPT to write an email on behalf of the manager with no other information than that—we just want it to write an email.

Prompt 4-3. Asking ChatGPT to write an email

```
Write an email from the perspective of a manager sending an
email to their team stating that they need to change over to
2-factor authentication to sign in to their emails.
```

In Response 4-3, we can see how ChatGPT responds to this very simplistic prompt.

Response 4-3. ChatGPT creates a very AI-sounding email

Subject: Important Update: Transition to Two-Factor Authentication for Email Access

Dear Team,

I hope this email finds you well. To further enhance the security of our company's communication systems and protect sensitive information, we are implementing two-factor authentication (2FA) for email access. This change is essential to ensure our accounts remain secure against potential cyber threats.

What You Need to Know

Starting [insert effective date], all employees will need to use 2FA to sign in to their work emails. This means that, in addition to your password, you will also be required to enter a verification code sent to your phone or email, or use an authenticator app.

Next Steps

1. Set Up 2FA:
   - Follow the instructions in the attached guide to set up 2FA for your account.
   - If you encounter any issues during the setup, our IT team will be available to assist you.
2. Test Your Access:
   - Once you've set up 2FA, test your access to ensure everything works smoothly.

3. Complete Setup by [insert deadline]:
   • Please ensure 2FA is fully set up before the
     effective date.

Why This Change?

The increasing sophistication of cyberattacks makes it critical
to adopt stronger security measures. 2FA significantly reduces
the risk of unauthorized access and safeguards our company data
and systems.

Your cooperation in this transition is greatly appreciated.
Please don't hesitate to reach out to me or the IT team at
[contact information] if you have any questions or concerns.

Thank you for your attention to this important update and for
helping us maintain a secure working environment.

Best regards,
[Your Name]
[Your Job Title]
[Your Contact Information]

Response 4-3 shows the kind of response that makes a lot of people
cringe when they think of using ChatGPT to write anything. And this is
simply because no other information has been provided other than the
task we wanted accomplished. This response essentially has no tone, style,
or anything unique about it other than the fact that ChatGPT wrote it.
And it shows. So let's find out how we can avoid this kind of response and
actually get something usable from ChatGPT.

Let's Gather Information to Give to ChatGPT

So the reason why our first attempt at writing an email was a bust was because we didn't give ChatGPT any information about the boss themselves. If you're an admin writing an email to the rest of the company, one of the most important things that needs to be done is to match the tone and style of the person that you are authorized to represent. This is where it's helpful to teach ChatGPT what emails from management typically sound like in order to help the admin craft an email appropriately. So the first thing the admin should do is collect a few emails from the boss to give ChatGPT as reference.

Therefore, the first step is to gather information that can be used to train ChatGPT to create the response we're actually looking for. In order to do that, we're going to give ChatGPT examples. We're going to use some of the manager's previous messages as training material for ChatGPT (Listing 4-2).

Listing 4-2. Example emails from Anthony Davis, technical manager at InnovateTech Solutions

Subject: Amazing Partnership Opportunity with InnovateTech Solutions!

Dear Helen Miller at NexxGen Synergy,

Hope you're doing great! I'm Anthony Davis, Technical Manager at InnovateTech Solutions. We talked during our meeting with the marketing team, but I wanted to formally introduce myself! I'm extremely excited about our teams working together. I think this partnership is bound to be awesome. How about a meeting next week to chat about helping our teams collaborate smoothly?

85

Catch you later,
Anthony Davis
Technical Manager at InnovateTech Solutions

—

Subject: You Belong at Our InnovateTech Summit Conference this Year!

Dear Mark Wilson,

I'm Anthony Davis at Innovate Tech and guess what? I've got the best news. We want YOU to be a keynote speaker at our big InnovateTech Summit 2024: Shaping the Future of Technology. We think your knowledge on security is amazing, we'd love for you to present. Let's make this happen.

Let's talk,
Anthony Davis
Technical Manager at InnovateTech Solutions

—

Subject: Team Meeting: Don't Be Late!

Hey Team,

Just a heads-up, we've got a team meeting today at 5pm. We'll talk about all the cool stuff we're doing. We'll also be looking at a few of our Q3 plans for the Marketing, Development, and IT teams. Let's go team!

Cheers,
Anthony Davis
Technical Manager at InnovateTech Solutions

Giving ChatGPT Training Material on the Manager's Writing Style so It Can Write an Email on Their Behalf to the Rest of the Company

Now that we have some emails from our imaginary boss, let's give ChatGPT these examples to help write an email to the rest of the company about a new security requirement. You can see our approach in Prompt 4-4.

Prompt 4-4. Asking ChatGPT to write emails in the voice of the technical manager of InnovateTech Solutions

You are an administrative assistant working under Anthony Davis at InnovateTech Solutions. You have to send an email to the team about switching to two-factor authentication for all sign in going forward. Here is an example of emails from Anthony:

\#\#\#

Subject: Amazing Partnership Opportunity with InnovateTech Solutions!

Dear Helen Miller at NexxGen Synergy,

Hope you're doing great! I'm Anthony Davis, Technical Manager at InnovateTech Solutions. We talked during our meeting with the marketing team, but I wanted to formally introduce myself! I'm extremely excited about our teams working together. I think this partnership is bound to be awesome. How about a meeting next week to chat about helping our teams collaborate smoothly?

Catch you later,
Anthony Davis
Technical Manager at InnovateTech Solutions

—

Subject: You Belong at Our InnovateTech Summit Conference this Year!

Dear Mark Wilson,

I'm Anthony Davis at Innovate Tech and guess what? I've got the best news. We want YOU to be a keynote speaker at our big InnovateTech Summit 2024: Shaping the Future of Technology. We think your knowledge on security is amazing, we'd love for you to present. Let's make this happen.

Let's talk,
Anthony Davis
Technical Manager at InnovateTech Solutions

—

Subject: Team Meeting: Don't Be Late!

Hey Team,

Just a heads-up, we've got a team meeting today at 5pm. We'll talk about all the cool stuff we're doing. We'll also be looking at a few of our Q3 plans for the Marketing, Development, and IT teams. Let's go team!

Cheers,
Anthony Davis
Technical Manager at InnovateTech Solutions

###

Write the email in the style of Anthony for the team.

In Response 4-4, we can see that when given training material, ChatGPT was able to represent the tone of management very well! Therefore, proper prompt engineering helps you avoid dry responses from ChatGPT when you're writing content.

Response 4-4. ChatGPT creates an email in the voice of the
technical manager of InnovateTech Solutions

Subject: Important Update: Two-Factor Authentication for All
Sign-Ins

Hey Team,

Hope you're all having a great day! I've got an important
update for you. To enhance our security and keep our data safe,
we're rolling out two-factor authentication (2FA) for all sign-
ins starting next week.

Why are we doing this? Simple - we want to make sure that our
accounts and sensitive information are protected. 2FA adds an
extra layer of security, ensuring that even if someone gets
hold of your password, they still can't access your account
without the second authentication factor.

Here's what you need to do:
1. Download the Authentication App: We recommend using Google
Authenticator or Authy.
2. Set Up 2FA: Follow the instructions we'll be sending out to
link your account with the authentication app.
3. Test It Out: Make sure it's working smoothly before the
switch next week.

Don't worry, it's a quick and easy process, and our IT team
will be available to help if you run into any issues.

Let's make sure we're all set and ready to go! This is a
crucial step to keep our InnovateTech Solutions family secure.

Thanks for your cooperation and let's keep rocking!

```
Cheers,
Anthony Davis
Technical Manager at InnovateTech Solutions
```

Now that we've seen what happens when we train ChatGPT on messages with a certain tone and writing style, let's see what happens when we tell ChatGPT a little bit about ourselves when we're asking for more personal advice.

Teaching ChatGPT to Make Task Management Easier

Let's step away from workplace-focused tasks and change gears for a moment. On an average, daily basis, ChatGPT can be used to increase productivity for anyone looking to save some time in their personal lives.

It can be hard for anyone to manage their time and calculate how long a task **actually** takes as opposed to how long they **think** it's going to take. People also tend to overestimate what they can do in a day and overwhelm themselves with a surplus of tasks with too little time. Instead of agonizing over being unproductive, let's teach ChatGPT what our plans for the day look like and ask it for help creating a manageable schedule. The prompt we used can be seen in Prompt 4-5.

Prompt 4-5. Giving ChatGPT our to-do list

```
I need help managing my time, and I can't tell how much I'll
actually be able to get done today. I get to work at 10am and
leave at 6pm. My commute is around half an hour. I'd prefer
to be done with tasks for the day by 8pm. Setting a priority
based on urgency and time needed to complete the task, make
a schedule for me, starting from 10am until 8pm. What can
```

I reasonably get done within that timeframe? Taking into consideration time constraints, travel time, and fatigue, what should I put off for tomorrow?

###

- Create a presentation on Pricing Strategies on Sales and Profit Margins
- Attend the marketing meeting with the CEO at 2pm
- Reply to an email from Maria
- Meal prep dinners for the week ahead
- Water the plants at home
- Go to the grocery store and pick up chicken breast for dinner
- Clean the oven
- Vacuum the living room
- Start a garden
- Edit 10 chapters of my new romance novel
- Finish reading the last 20 chapters of the meditation book I've been reading

In the example shown in Prompt 4-5, we've presented a hefty list of tasks with varying levels of importance. Let's see if ChatGPT can reasonably plot out a schedule that allows us to get the urgent tasks done and manage our time for the tasks that would reasonably take up a larger chunk of time than others.

Response 4-5. ChatGPT creates a timetable that prioritizes efficiency

Time Management Schedule for Today

Work Hours: 10am - 6pm (8 hours)
Commute: 30 minutes each way (1 hour total)
Preferred End Time for Tasks: 8pm

Prioritization of Tasks

1. High Priority (Must be done today)
   - Attend the marketing meeting with the CEO at 2pm
   - Reply to an email from Maria
   - Create a presentation on Pricing Strategies on Sales and Profit Margins

2. Medium Priority (Important but can be postponed if necessary)
   - Meal prep dinners for the week ahead
   - Water the plants at home
   - Go to the grocery store and pick up chicken breast for dinner

3. Low Priority (Can be postponed to tomorrow or later)
   - Clean the oven
   - Vacuum the living room
   - Start a garden
   - Edit 10 chapters of my new romance novel
   - Finish reading the last 20 chapters of the meditation book

Schedule

10:00am - 10:15am
- Settle in at work, check emails, and plan the day.

10:15am - 11:30am
- Reply to an email from Maria (15 minutes)
- Work on the presentation on Pricing Strategies (60 minutes)

11:30am - 12:30pm
- Continue working on the presentation on Pricing Strategies (60 minutes)

12:30pm - 1:00pm
- Lunch break

1:00pm - 2:00pm
- Finalize the presentation on Pricing Strategies (60 minutes)

2:00pm - 3:00pm
- Attend the marketing meeting with the CEO

3:00pm - 4:00pm
- Review and refine the presentation, handle any immediate follow-ups from the meeting (60 minutes)

4:00pm - 5:30pm
- Start working on editing the romance novel (90 minutes)

5:30pm - 6:00pm
- Finalize work tasks, check and reply to any remaining emails, prepare for the next workday

6:00pm - 6:30pm
- Commute home

6:30pm - 6:45pm
- Water the plants at home (15 minutes)

6:45pm - 7:15pm
- Go to the grocery store and pick up chicken breast for dinner (30 minutes)

7:15pm - 8:00pm
- Meal prep dinners for the week ahead (45 minutes)

Tasks to Postpone for Tomorrow:
- Clean the oven
- Vacuum the living room
- Start a garden
- Edit 10 chapters of my new romance novel (continue tomorrow)
- Finish reading the last 20 chapters of the meditation book

This schedule prioritizes high-importance tasks and ensures that essential work and personal tasks are completed. Tasks that are not urgent or crucial can be deferred to the following day to avoid overburdening yourself and to allow for a balanced workload.

Here, ChatGPT has planned out a timetable and to-do list that prioritizes important tasks and tasks that don't require a lot of time to accomplish. It delegates time for breaks and suggests which tasks can be put off, encouraging a balanced schedule that doesn't cause burnout.

Now, up until now, we've been training ChatGPT on information we already have, like Instagram posts, emails, and a to-do list. Teaching ChatGPT this kind of information is good because it ensures that ChatGPT doesn't *hallucinate* details, as we mentioned in Chapter 1. In other words, we're making sure ChatGPT doesn't make up information that doesn't exist, such as a nonexistent restaurant for an Instagram post, a false company name in an email, or a list of tasks you have no means to accomplish. However, because ChatGPT can be instructed to search the web for information, we can train ChatGPT based upon real findings— meaning it can't possibly hallucinate because the facts are readily available **and** current.

Teaching ChatGPT Up-to-Date Information and Training It to Create Blog Posts

Using ChatGPT's web search ability is a powerful capability and is essential to creating content that remains accurate and up to date. This is also important because ChatGPT cites its sources in its response, which allows you to fact-check where ChatGPT got its information. For example, you wouldn't want ChatGPT writing an article on politics based upon information it found from a gossip column.

That said, we're going to be putting the web search feature to use by asking ChatGPT to help us write a blog post for an environmentally conscious blog called The Happy Planet. First, we're going to do what we did before with the example of teaching ChatGPT to write in the style of a tech manager. We're going to offer ChatGPT examples of previous blog posts written by this fictional blogger.

Writing in the Style of Emily Carlton, Owner of The Happy Planet Blog

Let's present ChatGPT with an example of how a blog post from the author of The Happy Planet typically looks like. Listing 4-3 is an example of one of her posts on reducing the carbon footprint.

Listing 4-3. An example post from the eco-conscious blog, The Happy Planet

Hey eco-friends!

Welcome back to our deep dive on how to shrink that carbon footprint of ours. This time, we're going even further into the details, sharing more tips, tricks, and the latest innovations in sustainability. Whether you're a seasoned green warrior or just starting your journey, there's something here for everyone. Let's dive in!

What Exactly is a Carbon Footprint?

Before we jump into the tips, let's revisit what a carbon footprint is. It's the total amount of greenhouse gases, especially CO_2, that we produce through our daily activities. This includes everything from the food we eat to how we travel and the energy we use at home.

Globally, the average carbon footprint per person is about 4 tons of CO_2 annually, but in the U.S., it's around 16 tons per person. To meet climate goals, we need to bring this down to under 2 tons by 2050. So, how can we get there?

Actionable Tips for Everyday Life

1. Transportation: Pedal and Public Power

One of the biggest contributors to our carbon footprints is how we get around. Cars, especially those running on fossil fuels, emit a ton of CO_2. Here are some alternatives:
- Bike or Walk: Not only will you cut down on emissions, but you'll also get a good workout in. It's a win-win!
- Public Transport: Buses, trains, and subways are more eco-friendly than individual car rides. Plus, they often save you money and reduce traffic congestion.
- Carpool: If driving is a must, share the ride with friends or coworkers. It reduces the number of vehicles on the road and splits fuel costs.

2. Eat More Plants

Food production, especially meat, is a massive source of greenhouse gases. By adopting a more plant-based diet, you can significantly reduce your carbon footprint:
- Meatless Mondays: Start by cutting out meat one day a week. Gradually increase as you find more delicious plant-based recipes.
- Local and Seasonal: Buy locally-produced and seasonal foods to cut down on the transportation emissions associated with your meals.

3. Energy Efficiency at Home

Simple changes at home can lead to big savings in both energy and money:
- LED Bulbs: These use a fraction of the energy compared to traditional incandescent bulbs and last much longer.
- Smart Thermostats: These devices optimize heating and cooling, reducing unnecessary energy use.
- Unplug Devices: Many electronics draw power even when they're off. Unplugging devices when not in use can prevent this "phantom" energy loss.

4. Reduce, Reuse, Recycle

Minimizing waste is crucial for cutting emissions:
- Recycle Properly: Follow local recycling guidelines to ensure materials are processed correctly.
- Upcycle: Get creative with old items instead of tossing them. Turn an old t-shirt into a shopping bag or use jars as storage containers.

- Buy Less: Consider if you really need new items or if second-hand options could suffice. This is particularly important for clothing, where fast fashion is a significant environmental burden.

5. Water Conservation

Using less water reduces the energy required to pump, heat, and treat it:
- Shorter Showers: Aim for 5-minute showers.
- Fix Leaks: A dripping tap can waste gallons of water a day.
- Water-Efficient Fixtures: Install low-flow shower heads and toilets to save water without sacrificing performance.

6. Support Green Businesses

Support companies that prioritize sustainability:
- Eco-Friendly Products: Opt for products from brands that use sustainable practices, like biodegradable packaging and cruelty-free ingredients.
- Carbon Offsets: Some businesses invest in projects that offset their carbon emissions, such as tree planting or renewable energy projects.

Latest Innovations in Sustainability

Paris 2024 Olympics: A Case Study in Sustainability

The upcoming Paris Olympics are setting new standards in sustainability. The event aims to halve the carbon footprint compared to previous games by using 100% renewable energy, minimizing new construction, and promoting plant-based diets for athletes and spectators.

Circular Economy

The circular economy is gaining traction, emphasizing the reuse and recycling of materials to minimize waste. This approach not only reduces emissions but also creates economic opportunities. Companies are redesigning products to be more durable and recyclable, supporting a more sustainable model of consumption.

Biodiversity and Nature-Positive Initiatives

Protecting biodiversity is now seen as crucial for combating climate change. Initiatives include establishing protected areas, restoring ecosystems, and promoting sustainable agriculture. The goal is to ensure there is more nature by 2030 than there is today.

Conclusion

Reducing your carbon footprint is all about making mindful, sustainable choices. Start small, be consistent, and remember that every little bit helps. Whether it's biking to work, eating more plants, or supporting green businesses, your actions can make a huge difference.

Let's keep our adventures eco-friendly and our planet thriving. Together, we can create a healthier, greener world for ourselves and future generations. Stay green and keep shining! Feel free to reach out with your own tips and experiences. Let's keep the conversation going and inspire each other to live more sustainably!

Sources:

1. Our World in Data. CO₂ emissions. Retrieved from [Our World in Data](https://ourworldindata.org).
2. Nature.org. What is your carbon footprint? Retrieved from [Nature.org](https://www.nature.org).
3. World Economic Forum. How Paris Olympics 2024 could be the 'greenest-ever' games — https://www.weforum.org
4. IBM Blog. Sustainability trends: 5 issues to watch in 2024 — https://www.ibm.com
5. Nature.com. Thousands of studies reveal the best ways to cut your carbon footprint — https://www.nature.com
6. GreenCitizen.com. 21 Practical Ways on How to Reduce Your Carbon Footprint —https://www.greencitizen.com

Asking ChatGPT to Write a Blog Post About Sustainable Living

With the latest post from Emily Carlton as inspiration, we're going to ask ChatGPT to write about a different subject. We want ChatGPT to write about sustainable fashion, a topic that has been talked about often in the environmentally aware community.

Note   For the sake of brevity, we're indicating where we have given ChatGPT the contents of the blog post shown in Listing 4-3 with "[example goes here]" in Prompt 4-6.

Prompt 4-6. Asking ChatGPT to write a blog post about fast fashion

You are a blogger for the site The Happy Planet, a blog
dedicated to educating the people on the environment and
contributing factors to the climate crisis, as well as
providing tips on how to live more sustainably. You're to write
a blog post in the voice of the main author, Emily Carlton.
Here's an example of her latest post:

###

[example goes here]

###

In Emily's voice and style, write a new blog post on the
following subject and cover the following bullet points. Use
web search to find the latest news and resources, and cite your
sources. Write in prose:

Blog post title: Sustainable Fashion: How to Build an Eco-
Friendly Wardrobe
- Research sustainable fashion brands and their practices.
- Give examples of fast-fashion brands that would be best
  to avoid.
- Provide tips on how to shop ethically and sustainably.

###

To help with the process of creating the post, indicate where
it would be prudent to add photos to help break up text and
maintain viewer interest. Also suggest what photos would be
appropriate for these sections.

Now let's see how ChatGPT handles the task, as you'll see in
Response 4-6.

Response 4-6. ChatGPT creates a great post on environmentally conscious shopping habits

Sustainable Fashion: How to Build an Eco-Friendly Wardrobe

Hey eco-friends!

Welcome back to The Happy Planet! Today, we're diving into the world of sustainable fashion. Building an eco-friendly wardrobe can seem daunting, but with a bit of knowledge and some strategic choices, you can make a positive impact on the environment while looking fabulous. Let's explore how to shop sustainably, which brands to support, and which fast-fashion giants to avoid.

Sustainable Fashion Brands and Their Practices

When it comes to sustainable fashion, transparency and ethical practices are key. Here are some top brands leading the charge:

1. Patagonia: Known for its commitment to the environment, Patagonia uses sustainable materials like organic cotton and recycled wool. They also offer a repair program and buy back their own products to reduce waste.

2. Everlane: This brand focuses on radical transparency, eco-friendly materials, and ethical manufacturing processes. Everlane's commitment to sustainability includes eliminating virgin plastic from their supply chain and working towards net-zero emissions.

3. tentree: For every item purchased, tentree plants ten trees, aiming for a billion by 2030. Their products are made from eco-conscious fabrics like Tencel, hemp, and organic cotton.

4. Girlfriend Collective: Specializing in activewear, this brand uses recycled materials like fishing nets and water bottles. They also offer a recycling program for their products.

5. Reformation: Reformation combines stylish designs with sustainable practices. They use recycled and upcycled materials, are 100% carbon neutral, and invest in charitable programs.

Fast-Fashion Brands to Avoid

Fast fashion is notorious for its negative impact on the environment due to unsustainable practices and poor labor conditions. Here are a few brands to be cautious of:

1. Shein: Known for its extremely low prices, Shein has faced criticism for its lack of transparency and questionable labor practices.
2. Zara: Despite some efforts towards sustainability, Zara's business model of rapid production and frequent new collections contributes significantly to fashion waste.
3. H&M: Although H&M has a Conscious Collection, the brand's overall business model still relies heavily on high turnover and mass production, leading to significant environmental impacts.

Tips for Shopping Ethically and Sustainably

1. Buy Less, Choose Well: Invest in high-quality pieces that you truly love and will wear for years. This reduces the need for frequent purchases and helps you build a timeless wardrobe.

2. Second-Hand Shopping: Thrift stores, consignment shops, and online platforms like Poshmark and Depop offer great sustainable options. Buying second-hand extends the life of clothing and reduces waste.

3. Look for Certifications: Certifications like GOTS (Global Organic Textile Standard), Fair Trade, and B Corp indicate that a brand adheres to high environmental and ethical standards.

4. Support Local and Small Businesses: Many local and small brands focus on sustainable practices and craftsmanship. Supporting them can reduce the carbon footprint associated with shipping and promote ethical labor practices.

5. Care for Your Clothes: Proper care can extend the life of your clothes. Wash less frequently, use cold water, and avoid tumble drying to reduce energy use and fabric wear.

Visual Break: Add Photos

- Sustainable Fashion Brands: Display images of popular sustainable fashion items from brands like Patagonia, Everlane, and Girlfriend Collective.
- Fast-Fashion Brands to Avoid: Use a simple graphic to highlight brands like Shein, Zara, and H&M.
- Shopping Tips: Include photos of second-hand shops, clothing with certification labels, and local boutiques.

Conclusion

Transitioning to a sustainable wardrobe is a journey, but every step counts. By supporting ethical brands, making mindful purchases, and caring for your clothes, you can significantly reduce your environmental footprint. Let's continue to inspire each other towards a greener, more sustainable future. Feel free to share your sustainable fashion tips and favorite brands in the comments below!

Stay green and keep shining!

Sources:
1. 99 Sustainable Clothing Brands By Budget — https://www. thegoodtrade.com
2. Affordable Ethical and Sustainable Fashion Brands — https:// www.sustainably-chic.com
3. 31 Best Sustainable Clothing Brands — https://www.rd.com
4. 26 Sustainable Clothing Brands Making Garments for Good — https://www.shopify.com

Response 4-6 shows ChatGPT creating a very thorough post that not only suits the tone of the blog but also provides links that allow us to back-check its sources. ChatGPT even provides input on where images could be placed on the blog post to make it a more engaging read.

Conclusion

This chapter covered how ChatGPT can be taught to create more usable content for a variety of circumstances. We saw how prompting ChatGPT without additional information than the request itself results in very dry responses and that the simple way to solve the problem is to train ChatGPT on the kind of content we want it to produce. We also saw how ChatGPT can make Internet searches to ensure its information is factual and recent.

In the next chapter, we're going to carry on with what we've learned about training ChatGPT for tone and style as we use the Canvas feature to write side by side with ChatGPT for a thesis.

CHAPTER 5

Using ChatGPT Canvas to Help You Write Long-Form Content

For many students, writing a paper or thesis is a very daunting task and can leave some with a sense of paralysis that prevents them from even starting. Procrastination sets in, time passes, and the end result is a paper completed in the middle of the night by a student hopped up on Monster Energy drinks the day before the deadline. We've all seen this story before.

In Chapter 4, you saw how we used ChatGPT to write emails and create a list of tasks or a to-do list. We stretched its capabilities to create long-form content by asking it to write a blog post—which it did—but writing longer content like this is more inconvenient when you're simply having a back-and-forth conversation with ChatGPT. If you have small changes to make or want to make changes yourself, it's not easy to do that in this "chat window" format that is the default when you're using ChatGPT. Also, ChatGPT rarely writes anything longer than a few hundred words in its default mode.

© Lydia Evelyn 2025
L. Evelyn, *Making ChatGPT Work for You*, https://doi.org/10.1007/979-8-8688-1445-7_5

Well, ChatGPT's Canvas feature solves these problems by making ChatGPT a collaborator in your content creation, particularly when you're asking it to write longer bodies of text. In this chapter, we're going to look at how this feature works.

Using Canvas to Write a Thesis with ChatGPT

Up until now, we've been having a back-and-forth conversation with ChatGPT as we're already accustomed to at this point. However, in our next prompts, we're going to instruct ChatGPT to start using the Canvas feature. In this case, we're using Canvas to help us write an introduction to our thesis. Additionally, we're going to be asking ChatGPT to look for sources, just like we did in Chapter 4. This is an academic paper, after all. Therefore, we need ChatGPT to provide sources so we can see for ourselves where it got its information from. For our example, we're going to be writing an introduction to a thesis on how J.R.R. Tolkien's use of mythology and folklore influenced the development of modern fantasy literature.

Note   It's important to remember that different universities will have different rules on how much ChatGPT can be used to help with homework. Above all, university rules should be followed. This chapter will provide details that can be used to help write academic papers, but it's up to a student's discretion on how much advice to take from this chapter.

What will be covered in this chapter:

- What is Canvas and how does it work?

- Using ChatGPT to pick a topic for a thesis and get out of decision paralysis

- Choosing a focus for a thesis on J.R.R. Tolkien

- Using ChatGPT to help write the introduction of a thesis on J.R.R. Tolkien's effect on modern depictions of fantasy in literature and film

Now, let's answer the obvious question.

What Is Canvas?

Simply put, Canvas is a feature within ChatGPT that allows you to write alongside ChatGPT on long-form content. Think of it like your own personal collaborator. In order to leverage the Canvas feature, we have to switch to the Canvas "model" from our typical ChatGPT window. Figure 5-1 shows you how.

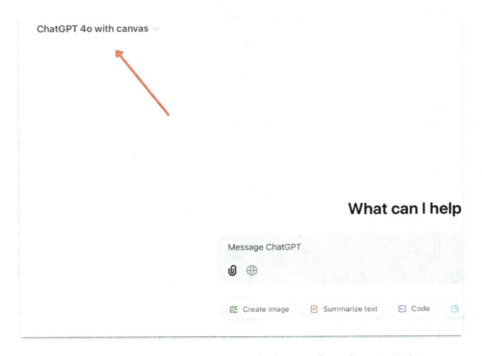

Figure 5-1. *Switching the model you're using for ChatGPT Canvas*

By clicking the drop-down on the top-left corner of any ChatGPT conversation window, you can select any model that has the Canvas feature to get started. What this will do is allow ChatGPT to decide if what you're asking for will require the creation of long-form content. Alternatively, you can tell ChatGPT to use the Canvas feature explicitly. So let's try out this exciting feature with Prompt 5-1.

Prompt 5-1. Getting started

```
Let's write an article together.
```

Now, you'll notice that we used a very simple prompt. And, as you might have expected, we got a very simple answer from ChatGPT (Response 5-1).

Response 5-1. ChatGPT answers, but without Canvas

Sure! What topic do you have in mind for the article? We could explore something related to your current projects, a particular interest, or even brainstorm some ideas together!

As you can see, though we said we were ready to write an article, we haven't made ChatGPT think we were ready to open the Canvas mode. This is because we've given it no details on what we want to write about, how long the article will be, or anything else. Cleverly, you'll notice that ChatGPT actually suggests the kinds of prompts we can use to help us get started. So, let's take its advice and explore a specific topic. I picked elephants because... well, I like elephants and watch way too many *National Geographic* documentaries. We used Prompt 5-2 to continue our conversation.

Prompt 5-2. Let's write a simple article on elephants

I want to write about the evolution of the Asian elephant!

Our topic is much more specific now. We've chosen to write about the evolution of the Asian elephant, a subject that's sure to go on for much longer than a few hundred words if used in an article. In Figure 5-2, you'll see that ChatGPT decided that this was a job for the Canvas feature.

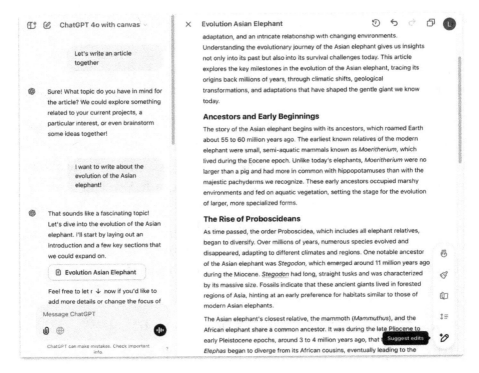

Figure 5-2. *The Canvas feature looks pretty impressive!*

Whoa, this looks very different from anything we've seen with ChatGPT so far, so let's slow down and understand what we're looking at.

Anatomy of a ChatGPT Canvas Window

If you haven't noticed already, the way we've used ChatGPT so far has been in the form of a conversation. We send a prompt; ChatGPT responds. If we wanted ChatGPT to modify its response in any way, we've had to make a suggestion, and then ChatGPT would send a new response with our requested changes. This is great, but can become cumbersome if you're writing more long-form content or if the changes you'd like ChatGPT to make only affect a small portion of the text it's sending you, not the entire response. Canvas fixes these problems in a few ways. Figure 5-3 shows a numbered guide to the key features Canvas offers that's different from the typical ChatGPT format we've been accustomed to thus far.

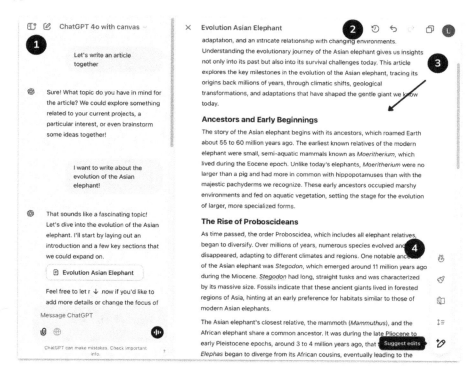

Figure 5-3. *Highlights of ChatGPT's Canvas features*

Let's break down each feature of ChatGPT's Canvas:

1. The chat window has become a panel that is pushed to the left side of the page. Here, we can hold a conversation with ChatGPT as normal. The difference here is that the **content we're asking ChatGPT to create** will be on the right side of the page.

2. These are buttons that allow you to undo, redo, copy the content, and restore a previous version of the content you're working on.

3. Here on the right side of the page, we have the **content we're creating with ChatGPT**. In this case, it's an article on the evolution of the Asian elephant. Unlike the typical responses we get from ChatGPT, this text can be manually edited by you **or** ChatGPT (Figure 5-4 shows what this looks like). What this means is you can make a minor change yourself or, in the chat window to the left, ask ChatGPT to make a change on your behalf. You can even highlight a portion of the text in this window and ask ChatGPT to make a specific change to that portion alone (again, see Figure 5-4 for details).

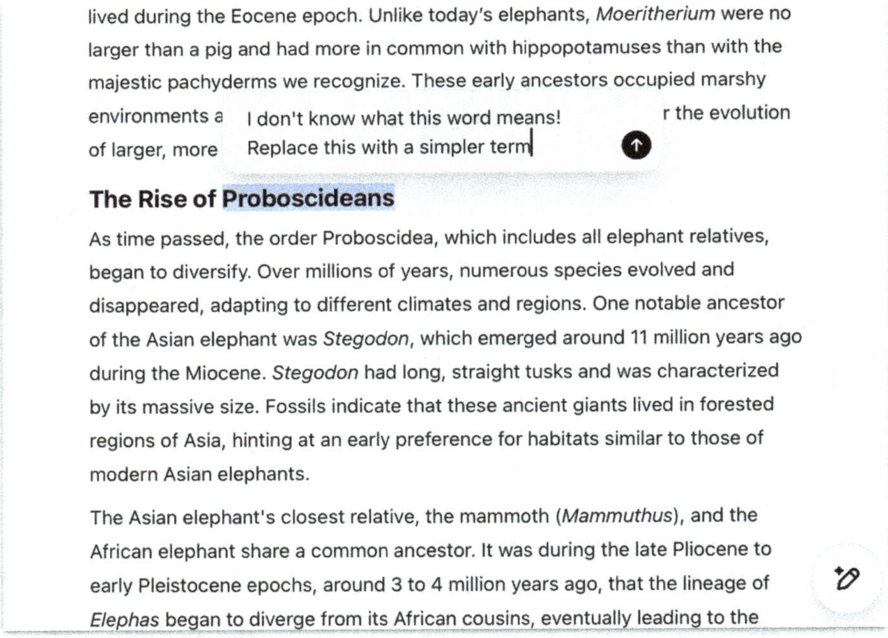

lived during the Eocene epoch. Unlike today's elephants, *Moeritherium* were no larger than a pig and had more in common with hippopotamuses than with the majestic pachyderms we recognize. These early ancestors occupied marshy environments a   I don't know what this word means!   r the evolution of larger, more   Replace this with a simpler term

The Rise of Proboscideans

As time passed, the order Proboscidea, which includes all elephant relatives, began to diversify. Over millions of years, numerous species evolved and disappeared, adapting to different climates and regions. One notable ancestor of the Asian elephant was *Stegodon*, which emerged around 11 million years ago during the Miocene. *Stegodon* had long, straight tusks and was characterized by its massive size. Fossils indicate that these ancient giants lived in forested regions of Asia, hinting at an early preference for habitats similar to those of modern Asian elephants.

The Asian elephant's closest relative, the mammoth (*Mammuthus*), and the African elephant share a common ancestor. It was during the late Pliocene to early Pleistocene epochs, around 3 to 4 million years ago, that the lineage of *Elephas* began to diverge from its African cousins, eventually leading to the

Figure 5-4. *You can highlight any portion of the text within the content window to make a small change*

4. This is the **Suggest edits** panel. It includes a list of commands you can tell ChatGPT to apply to the **content we're creating** as a whole. You can **add emojis**, which instructs ChatGPT to add emojis to the content wherever it deems suitable. You can **add final polish**, which essentially tells ChatGPT to "check its work" and ensure the quality of the content. You can adjust the **reading level**, which is very similar to the prompts we used in Chapter 1 to adjust the tone of an example science report from a fifth grade reading level to the academic tone of an English paper. Finally, you can **adjust the length** of the document, which is pretty self-explanatory. Adjusting this parameter is a way to instruct ChatGPT to either elaborate on what the content covers to make it longer or condense the content to make it shorter.

If you've ever collaborated with someone else on a Word document or Google Docs file, you'll see that the process of using the Canvas feature is very similar. The Canvas window becomes a text editor that you can type in freely and even make text bold, italicized, or underlined. You can also create headers!

So, now that we know what Canvas is and what it does, let's ask ChatGPT to help us write the introduction to a thesis with the help of the Canvas feature. But first, let's start by brainstorming.

Picking a Topic for Our Thesis

Before we start using Canvas for writing our thesis, we're going to start the process from the beginning and get ideas from ChatGPT on what to write about in the first place.

Let's say you need to write a paper for your English Literature class based on influential European authors that have made major impacts on American literature. But let's take a step back. You may not even know what you want to write about. Let's release ourselves from decision paralysis and get ChatGPT to offer us some options to get us started. We used Prompt 5-3.

Note   Keep in mind that we're going to be using the Canvas model for the rest of this chapter.

Prompt 5-3. Conceptualizing a thesis

```
I need to write a thesis for my English Literature class on
European authors and works of literature that have greatly
impacted today's most influential novels, films, and the
culture as a whole. Can you suggest some authors and/or titles
for this project?
```

This time, our prompt is very specific. We've provided ChatGPT with as many details as possible on what we're trying to create so that ChatGPT's answer will be relevant to what we want. Remember how just asking for an article didn't get us a helpful response? That's why we're making our request clear and detailed. The more detailed you are in a prompt, the better. Let's see what suggestions we get from ChatGPT. You'll see the response we got in Response 5-3.

Response 5-3. ChatGPT suggests topics for an English literature thesis

```
That sounds like a fascinating thesis topic! Here are some
European authors and works that have had a significant impact
on modern literature, film, and culture:
```

1. William Shakespeare (England)
- **Works:** *Hamlet, Macbeth, Romeo and Juliet*
- **Impact**: Shakespeare's plays have influenced countless novels, films, and cultural references. His exploration of human nature, complex characters, and enduring themes are foundational in both drama and storytelling.

2. Jane Austen (England)
- **Works:** *Pride and Prejudice, Emma*
- **Impact:** Austen's novels laid the groundwork for the modern romantic comedy genre and have inspired numerous adaptations. Her portrayal of social norms, character development, and wit are still echoed in contemporary literature and media.

3. Mary Shelley (England)
- **Work:** *Frankenstein*
- **Impact:** Often regarded as the first work of science fiction, *Frankenstein* has greatly influenced the sci-fi genre, themes of ethics in scientific discovery, and the "mad scientist" trope seen in film and literature today.

4. Victor Hugo (France)
- **Works:** *Les Misérables, The Hunchback of Notre-Dame*
- **Impact:** Hugo's work deals with social justice, moral struggles, and the plight of the underprivileged. *Les Misérables*, in particular, has been adapted into many films and musicals, impacting the portrayal of revolution and human struggle.

5. Fyodor Dostoevsky (Russia)
- **Works:** *Crime and Punishment, The Brothers Karamazov*
- **Impact:** Dostoevsky's psychological exploration of characters and philosophical themes has influenced modern

existential literature and film. His insights into
morality and human suffering are seen in both novels and
movies tackling deep philosophical issues.

6. Franz Kafka (Austria-Hungary/Czechoslovakia)
- **Works:** *The Metamorphosis, The Trial*
- **Impact:** Kafka's themes of alienation, bureaucratic
  absurdity, and existential anxiety are seen in both
  literature and films that depict the struggles of the
  individual against incomprehensible systems.

7. Charlotte and Emily Brontë (England)
- **Works:** *Jane Eyre* (Charlotte), *Wuthering Heights* (Emily)
- **Impact:** The Brontë sisters' novels have had a lasting
  influence on gothic literature and themes of passionate,
  tragic love, which can be seen in many modern novels
  and films.

8. Leo Tolstoy (Russia)
- **Works:** *War and Peace, Anna Karenina*
- **Impact:** Tolstoy's exploration of human experiences,
  relationships, and society has influenced epic
  storytelling in novels and film, particularly those
  dealing with historical settings and personal struggles
  amid societal changes.

9. Charles Dickens (England)
- **Works:** *Great Expectations, Oliver Twist, A Tale of Two Cities*
- **Impact:** Dickens' focus on social issues, especially those
  affecting the poor, as well as his memorable characters,
  has inspired countless adaptations in film and literature.

10. Jules Verne (France)
- **Works:** *Twenty Thousand Leagues Under the Sea, Around the
  World in Eighty Days*

- **Impact:** Verne is considered a pioneer of science fiction and adventure storytelling. His works have influenced modern sci-fi, adventure films, and even technological innovations.

11. George Orwell (England)
- **Work:** *1984*
- **Impact:** Orwell's dystopian vision has profoundly shaped the way we perceive surveillance, totalitarianism, and social control. His influence is seen in films like *The Matrix*, *V for Vendetta*, and numerous dystopian novels.

12. J.R.R. Tolkien (England)
- **Works:** *The Lord of the Rings* trilogy, *The Hobbit*
- **Impact:** Tolkien's works have defined the modern fantasy genre and have been a major influence on films like *The Lord of the Rings* and series like *Game of Thrones*. His creation of a richly detailed secondary world set the standard for modern fantasy literature.

These authors and their works provide a broad perspective on how European literature has shaped current storytelling, across themes such as social justice, romantic ideals, human psychology, science fiction, fantasy, and existentialism. Any combination of these influences could make for a compelling thesis on how these foundational works have impacted today's novels, films, and even broader cultural trends.

Let me know if you need help narrowing down your focus or developing your thesis statement!

Response 5-3 is incredibly detailed and helpful. For a student that doesn't even know where to start, this is enough to get plenty of ideas flowing. ChatGPT lists a variety of topics and authors. It even proposes various approaches that would favor a well-written paper. So, now that we have a list of options, let's pick one for our thesis.

Choosing the Focus of Our Thesis

Out of all of the options ChatGPT gave us, we're going to pick J.R.R. Tolkien and his works as the focus of our thesis. Our goal is to write about the effect of a European author on American literature, and Tolkien is credited for having inspired prominent books and movies, such as *Game of Thrones* and *The Lord of the Rings* movies. So let's get some ideas on how we can use J.R.R. Tolkien as a topic for this thesis. We used Prompt 5-4.

Prompt 5-4. Explaining the goal of our thesis

I want the focus on my thesis to be about JRR Tolkien. Give me ideas for a research question that I could use as the subject of my thesis.

This prompt is simple, yet specific. We're using ChatGPT for every step of the process, allowing us to have all of the details for our thesis prepared before we even start writing it. We can see in Response 5-4 that ChatGPT comes back with fantastic suggestions.

Response 5-4. A plethora of options for a thesis on J.R.R. Tolkien

Focusing on J.R.R. Tolkien offers a wealth of fascinating directions for your thesis. Here are some research question ideas that could serve as the subject of your thesis, each exploring a different aspect of Tolkien's influence and legacy:

1. How did J.R.R. Tolkien's use of mythology and folklore influence the development of modern fantasy literature?

120

- This question could explore the mythological roots of *The Lord of the Rings* and *The Hobbit*, and how Tolkien's use of Norse, Celtic, and Christian mythology shaped the conventions of the fantasy genre.

2. In what ways did J.R.R. Tolkien's experiences in World War I influence the themes of heroism and loss in *The Lord of the Rings*?

- This topic could focus on how Tolkien's personal history shaped his depiction of war, camaraderie, and the costs of power, drawing parallels between real-world history and Middle-earth's conflicts.

3. How does J.R.R. Tolkien's world-building in *The Lord of the Rings* and *The Silmarillion* redefine the concept of a "secondary world," and why is this important to the fantasy genre today?

- This question allows you to analyze Tolkien's approach to world-building in depth, focusing on how his attention to history, languages, and cultures has set the standard for immersive storytelling in fantasy.

4. To what extent did J.R.R. Tolkien's linguistic expertise influence his storytelling, and how has this shaped the development of modern constructed languages in fantasy literature and media?

- Tolkien was a philologist, and his passion for languages heavily influenced the creation of Middle-earth. This topic would allow you to explore how his linguistic creations have influenced other authors and media, including constructed languages like Dothraki from *Game of Thrones* or Klingon from *Star Trek*.

5. How does J.R.R. Tolkien use the concept of power and corruption in *The Lord of the Rings,* and what does it reveal about his views on authority and moral responsibility?

- This question could examine the portrayal of the One Ring as a symbol of corrupting power and explore how Tolkien's narrative reflects his views on the nature of power and the danger it poses to individuals and society.

6. How did J.R.R. Tolkien's portrayal of nature and industrialization in *The Lord of the Rings* reflect the socio-political concerns of his time?

- This thesis could focus on Tolkien's depiction of pastoral landscapes versus industrial encroachment, exploring his concerns about environmental destruction, which was especially poignant in the post-industrial revolution and post-war era.

7. What role do female characters play in J.R.R. Tolkien's Middle-earth, and how has his depiction of women influenced later representations of gender in fantasy literature?

- This topic allows you to explore characters like Éowyn, Galadriel, and Arwen, examining their roles, their impact on the plot, and how their portrayal has influenced female representation in later fantasy works.

8. How does J.R.R. Tolkien's concept of "eucatastrophe" function in *The Lord of the Rings*, and what significance does it hold for the fantasy genre as a whole?

- "Eucatastrophe" is a term Tolkien coined to describe a sudden, joyous turn of events. This question would allow you to analyze how this concept is employed in his works and why it has become a key component of the emotional impact in modern fantasy narratives.

9. How did J.R.R. Tolkien's portrayal of the heroic journey influence Joseph Campbell's *Hero with a Thousand Faces* and the subsequent rise of the hero's journey in modern storytelling?

- This question would allow you to explore the similarities between Tolkien's works and Campbell's concept of the monomyth, analyzing Tolkien's influence on the popularization of the heroic journey in books, films, and video games.

10. What is the significance of fellowship and community in J.R.R. Tolkien's *The Lord of the Rings*, and how does it reflect the author's perspective on unity and friendship during times of crisis?
- This thesis could analyze the dynamics within the Fellowship of the Ring and other relationships throughout the story, exploring themes of solidarity, sacrifice, and companionship as a response to overwhelming adversity.

These research questions could each serve as a strong foundation for a deep and meaningful exploration of Tolkien's work and legacy. Let me know if any of these resonate with you, or if you need help developing the question further!

ChatGPT has offered a wealth of options for our thesis in Response 5-4, and at this point a student would be spoiled for choice. This is certainly enough to have something to start with for someone who is completely stuck. The great thing about using ChatGPT is that it can provide you with as much ammunition as you need to get started or supply you with information and ideas to fill in the blanks when you've hit a wall.

Using ChatGPT to Help Us Write an Outline

Now, we could really pick any of the topics ChatGPT offered us; however, let's go with writing about how J.R.R. Tolkien's use of mythology and folklore influenced the development of modern fantasy literature.

Let's ask ChatGPT to help us write an outline (Prompt 5-5).

Prompt 5-5. Using ChatGPT to help write an outline for our thesis on J.R.R. Tolkien

I'm thinking about writing my thesis on how J.R.R. Tolkien's use of mythology and folklore influenced the development of modern fantasy literature. Help me write an outline.

Writing an outline is one of the most important steps to writing a thesis, and getting help from ChatGPT on this step is incredibly convenient. Response 5-5 shows the response we got from ChatGPT.

Response 5-5. ChatGPT creates an outline based on our request

Certainly! Here's an outline that will help structure your thesis on how J.R.R. Tolkien's use of mythology and folklore influenced the development of modern fantasy literature:

Thesis Title:
"The Mythical Foundations of Modern Fantasy: J.R.R. Tolkien's Influence through Mythology and Folklore"

I. Introduction
1. Background and Context
  - Brief introduction to J.R.R. Tolkien and his major works (*The Lord of the Rings, The Hobbit, The Silmarillion*).
  - Overview of modern fantasy literature and its growing cultural impact.

2. Research Question
  - How did J.R.R. Tolkien's use of mythology and folklore influence the development of modern fantasy literature?

3. Thesis Statement
- Tolkien's extensive use of mythological and folkloric elements not only shaped the narrative structure of his own works but also created a foundation for the themes, tropes, and conventions of the modern fantasy genre.

4. Outline of Approach
- Brief outline of the key sections and main points.

II. Mythological Foundations in Tolkien's Works
1. Norse Mythology
- *Connection to Norse Myths:* Discuss the parallels between Tolkien's characters (e.g., Gandalf and Odin) and Norse gods.
- *Influence on Themes:* Themes of heroism, fate, and valor similar to *The Prose Edda* and other Norse sagas.

2. Celtic and Arthurian Legend
- *Mythic Archetypes:* Explore how Tolkien borrowed motifs from Celtic mythology, such as magical swords (Glamdring, Sting).
- *Arthurian Influence:* The notion of a king who returns and the establishment of a rightful kingdom (echoes in Aragorn's story).

3. Classical Mythology
- *Greek and Roman Influences:* Discussion on Tolkien's adaptation of classical mythology motifs like epic journeys and tragic heroes.
- *Parallels with Homeric Tradition:* Comparison between the Fellowship's journey and epic quests in works like *The Odyssey*.

4. Christian Mythology
- *Christian Symbolism:* Tolkien's blending of myth and Christian allegory, emphasizing themes of sacrifice, resurrection, and redemption.

III. Folklore Elements in Tolkien's Narrative
1. British and European Folktales
- *Hobbits and Rural England:* Examination of how Tolkien was inspired by rural folklore in creating the Shire and Hobbits as symbols of pastoral simplicity.
- *Mythical Creatures:* Analysis of familiar creatures from folklore (elves, dwarves, trolls) and how Tolkien redefined their roles in fantasy literature.

2. Fairy Tales and Storytelling Traditions
- *"Faërie" and Secondary Worlds:* The idea of "faërie" as Tolkien described in his essay "On Fairy-Stories" and its significance in his world-building.
- *Transformation of Folk Heroes:* Bilbo and Frodo as folk heroes, the reluctant adventurer trope derived from traditional folk tales.

IV. Tolkien's Influence on the Modern Fantasy Genre
1. World-Building and Secondary Worlds
- *Creation of Middle-earth:* Discuss Tolkien's approach to constructing a fully realized world with languages, history, and geography.
- *Influence on Later Authors:* How this depth of world-building became a model for future fantasy writers like George R.R. Martin, Robert Jordan, and Patrick Rothfuss.

2. Fantasy Tropes and Archetypes
- *The Hero's Journey:* Analysis of how Tolkien helped popularize the heroic quest structure that is now a staple in fantasy.
- *The Mentor and Fellowship:* Influence of characters like Gandalf (the wise mentor) and the concept of an assembled fellowship of diverse beings.

3. Tolkien's Enduring Legacy in Popular Culture
- *Impact on Literature and Media:* Examination of Tolkien's influence on contemporary literature, films (*The Lord of the Rings* adaptations), and video games.
- *Establishment of Genre Norms:* Tolkien's role in setting genre conventions, such as the battle between good and evil, ancient artifacts of power, and rich, epic storytelling.

V. Case Studies of Influenced Works
1. Literature Inspired by Tolkien
- *C.S. Lewis and The Chronicles of Narnia:* Brief discussion of Tolkien's influence on his friend Lewis.
- *George R.R. Martin's* A Song of Ice and Fire: Analysis of how Martin adapted and diverged from Tolkien's mythic elements.

2. Fantasy Films and Series
- *Peter Jackson's Adaptations:* How Tolkien's works translated into film and their impact on the portrayal of mythology in visual media.
- *The Influence on Popular Series:* Comparison to modern franchises like *The Witcher, Harry Potter,* and *Game of Thrones.*

VI. Critiques and Interpretations of Tolkien's Use of Mythology
1. Academic Critiques
   - *Myth vs. Originality:* Discussion of critiques that argue Tolkien's work is too derivative of mythological sources.
   - *Tolkien's Defense of Myth-Making:* Reference to Tolkien's view on myth as a means of uncovering universal truths (as argued in "On Fairy-Stories").

2. Cultural and Societal Impact
   - *Relevance of Myth Today:* Analysis of why mythological themes remain relevant and how Tolkien revitalized interest in these stories.

VII. Conclusion
1. Summary of Key Points
   - Recap of Tolkien's use of mythology and folklore, and its impact on the fantasy genre.

2. Thesis Restatement
   - Reinforce how Tolkien's integration of mythological and folkloric elements created a blueprint for the modern fantasy genre.

3. Final Thoughts
   - Reflection on Tolkien's enduring legacy and the timeless appeal of myth in shaping new stories.
   - Suggestion for further research on the influence of Tolkien's works across different cultures and genres.

This outline gives you a solid structure to explore Tolkien's mythological roots and his profound influence on modern fantasy literature. Feel free to modify or expand any sections to suit your needs. Let me know if you need help developing any part of this further or want to discuss specific points in more detail!

This is a very thorough and comprehensive outline, and it meets our needs for an outline that will guide our research and further work that will need to be done after receiving approval from professors. From here, we know what our thesis is about and we have an outline. At this point, we can start using Canvas to write the introduction to the thesis.

Using the Canvas Feature to Write the Introduction to a Thesis on J.R.R. Tolkien

In our prompt, we're explicitly telling ChatGPT to use the canvas feature. In some cases, ChatGPT may open Canvas automatically, but we're simply going to instruct it to do so here. And just as we've done thus far, we're going to be very specific on what we're asking ChatGPT to create for us (Prompt 5-6).

Prompt 5-6. Using direct instructions to activate Canvas

Open Canvas and help me write the introduction of my thesis. I need examples of how JRR Tolkien influenced American literature. Give me a few authors, films, and novels that were inspired by Tolkien. I also need details on Tolkien and his inspirations. I'd like sources for my thesis, so name some critical essays that I can use as reference.

As you can see in Figure 5-5, ChatGPT opens Canvas like it did when we asked for help writing an article about the Asian elephant.

Figure 5-5. *Instructing ChatGPT to open Canvas so that we can work together*

In Response 5-6, we have the complete text that ChatGPT wrote for the thesis.

Response 5-6. ChatGPT helps us write the introduction to our thesis

Introduction

J.R.R. Tolkien's contribution to modern literature cannot be overstated; his works, notably *The Lord of the Rings*, *The Hobbit*, and *The Silmarillion*, redefined the fantasy genre and established the mythic foundation for what we now recognize as modern epic fantasy. Tolkien, with his deep love for mythology and folklore, provided a meticulously crafted world rich with complex histories, languages, and archetypes that continue to resonate in popular culture today. His influence on American literature, film, and entertainment is evident in the numerous authors and filmmakers who have drawn inspiration from Middle-earth, borrowing themes, tropes, and narrative structures to create compelling works that speak to contemporary audiences.

American authors like George R.R. Martin, whose *A Song of Ice and Fire* series echoes the epic scale and character-driven complexity found in Tolkien's work, openly acknowledge the influence of Middle-earth on their writing. Similarly, Terry Brooks' *Shannara* series draws heavily on Tolkien's world-building style, utilizing a quest-driven narrative involving ancient magic and a battle between good and evil. Another notable example is Patrick Rothfuss, whose *The Kingkiller Chronicle* employs a sophisticated level of world-building and lyrical storytelling akin to Tolkien's creation of mythic histories. These American fantasy authors have taken Tolkien's blueprint and adapted it for new generations, incorporating modern sensibilities while maintaining the grandeur and depth characteristic of Tolkien's works.

Tolkien's influence extends beyond literature and deeply into cinema and popular culture. Peter Jackson's *The Lord of the Rings* and *The Hobbit* trilogies brought Tolkien's expansive world to an even broader audience, inspiring new waves of fantasy films and television series. American productions such as *The Chronicles of Narnia* (adapted from the works of Tolkien's contemporary and friend, C.S. Lewis) and HBO's *Game of Thrones* clearly reflect Tolkien's narrative techniques, focusing on epic conflicts, richly drawn characters, and moral complexity. Moreover, films like *Harry Potter* also carry Tolkien's legacy, with their magical worlds, mysterious artifacts, and coming-of-age narratives centered around the battle between good and evil.

Tolkien's inspiration stemmed from his deep academic background in languages and mythology. He was a philologist and a professor of Anglo-Saxon at Oxford, and his expertise in Old English, Norse mythology, and early European folklore directly influenced the creation of Middle-earth. For instance, the character of Gandalf is inspired by Odin, the Norse god, while the concept of the Elves draws from both Norse and Celtic traditions. Tolkien also integrated elements of Christian mythology, exploring themes of sacrifice, redemption, and the struggle against a pervasive evil. His famous essay, "On Fairy-Stories," articulates his belief in the power of myth and storytelling as essential tools for understanding human experience, highlighting his intention to create a "mythology for England."

For further exploration of Tolkien's mythological foundations and influence, several critical essays offer invaluable insight. Tom Shippey's J.R.R. *Tolkien: Author of the Century* provides a detailed analysis of Tolkien's literary methods, while Verlyn Flieger's *Splintered Light: Logos and Language in Tolkien's World* delves into Tolkien's use of language and myth. Additionally, Shippey's *The Road to Middle-earth* is an essential resource for understanding the academic and folkloric roots of Tolkien's works, connecting his influences to his writing. These texts serve as key references to explore the depth of Tolkien's mythology and its profound impact on both the fantasy genre and modern storytelling as a whole.

This response is very... dry. It doesn't at all have an academic tone usable in a thesis, and it most certainly sounds like a response from ChatGPT. Not only that, though it cites sources, it does not do so in a way that is appropriate for an academic paper. What went wrong?

Well, we didn't incorporate all of the strategies we learned in Chapter 3. We need to give ChatGPT an **example** of the kind of content we're trying to create. We're also going about it the wrong way by simply telling it what we want. Remember, ChatGPT is good at role-play. Rather than ask for its advice, we need to tell it to be the student. So let's try that again.

Giving ChatGPT an Example of What We Want and Using the "You Are a..." Prompt Approach

Very briefly, we're going to tap into a capability of ChatGPT we have not yet covered. In this case, we're going to *upload a file* to ChatGPT—specifically, a PDF. This is because we're going to reference an existing academic paper that is too large to fit into the character limit of the text window of a conversation with ChatGPT.

We're going to cover more of this capability later in Chapter 6. However, for the purposes of this chapter, understand that we're uploading a PDF of an academic paper on the influence of Shakespeare's existentialism by Charlotte Keys, entitled "Shakespeare's Existentialism." This is very similar to the process of uploading a photo to ChatGPT, as covered in Chapter 4.

As you can see in Prompt 5-7, we've already uploaded the PDF with the example of "Shakespeare's Existentialism" as reference.

Prompt 5-7. Giving ChatGPT a "you are a..." prompt to solicit a better response

You are an English literature student writing about Tolkien's influence on modern American literature and film. The research question is "How did J.R.R. Tolkien's use of mythology and folklore influence the development of modern fantasy literature?" Write the introduction of your thesis to match the tone and voice of the thesis I've uploaded as reference, including all of the details we've already discussed. Your thesis needs to sound more personal, yet professional. It needs to capture the reader's attention and allow professors to see how much you appreciate the unique perspective Tolkien adds to the world of literature. Also, cite references as is appropriate for an academic paper.

For reference, Listing 5-1 is a short sample from the PDF we uploaded to ChatGPT.

Listing 5-1. A short sample of Charlotte Keys's "Shakespeare's Existentialism"

SHAKESPEARE'S EXISTENTIALISM: AN INTRODUCTION
Shakespeare and existentialists share a special philosophical kinship: both are fascinated by how human beings live in the world, how they experience themselves, and how they interact with and respond to other people. Shakespeare's plays - and his tragedies in particular - are full of existentially painful and intense moments. Time and again, Shakespeare shows his interest in complex ontological and existential issues by presenting characters who experience themselves as divided, damaged, and even dissolved. Richard III's syntactically disjointed speech after his nightmare on the eve of battle

134

demonstrates Shakespeare's particular aptitude for dramatising the disintegration
of subjectivity:

Cold fearful drops stand on my trembling flesh.
What do I fear? Myself? There's none else by.
Richard loves Richard; that is, I am I.
Is there a murderer here? No. Yes, I am.
Then fly! What, from myself? Great reason. Why?
Lest I revenge. Myself upon myself?
Alack, I love myself. Wherefore? For any good
That I myself have done unto myself?
O no, alas, I rather hate myself
For hateful deeds committed by myself.
I am a villain. Yet I lie: I am not.
Fool, of thyself speak well. (Richard III, V.v.135-46)

'Richard loves Richard': the narcissistic embrace of self is a desperate attempt by Richard to repair this internal fracturing and become self-identical, but the tight rhetorical structure of the speech and the self-estranging use of the third person are literally and figuratively self-defeating. As Eric Langley notes, 'Paradoxically, it is the language of self-assertion or definition that unravels in reflective repetition.' An act of self-identification becomes an act of self-annihilation for Richard. Rather than reaffirming his identity, his oscillations - 'No. Yes, . . . I love myself . . . I rather hate myself . . . I am . . . I am not' - cancel each other out and only succeed in intensifying and exacerbating his traumatic experience of self-loss.

Credit: Charlotte Keys's "Shakespeare's Existentialism"

Now that we've demonstrated to ChatGPT what kind of writing style we're looking for, let's see how ChatGPT was able to improve with our improved prompt. You can see the response we got from ChatGPT in Response 5-7.

Response 5-7. ChatGPT provides a response that sounds like an actual thesis introduction

Introduction
Tolkien's Influence on Modern Fantasy Authors

J.R.R. Tolkien's contribution to modern literature is not just a testament to his extraordinary imagination, but also a reflection of his deep-rooted passion for mythology and folklore. His works, notably *The Lord of the Rings*, *The Hobbit*, and *The Silmarillion*, have redefined the fantasy genre and established the foundation for what we now recognize as modern epic fantasy (Shippey, 2000). For me, Tolkien's ability to weave mythic narratives into richly detailed worlds has always represented more than just stories of good versus evil; it is a tapestry of timeless archetypes, a way for us to connect to our collective past while dreaming of the future. His influence on American literature, film, and entertainment is unmistakable, and countless authors and filmmakers have drawn inspiration from the world of Middle-earth, borrowing themes, tropes, and narrative structures to craft their own compelling works (Flieger, 2002).

Tolkien's Influence on American Fantasy Authors

American authors like George R.R. Martin, whose *A Song of Ice and Fire* series echoes the epic scale and character-driven complexity found in Tolkien's work, openly acknowledge the influence of Middle-earth on their writing (Shippey, 2005).

Martin's world is one where moral ambiguity and intricate politics reign, but at its heart, the influence of Tolkien's mythic framework is evident in the sweeping landscapes and the deep, character-driven narratives. Similarly, Terry Brooks' *Shannara* series draws heavily on Tolkien's world-building style, employing a quest-driven narrative involving ancient magic and a battle between good and evil (Shippey, 2000). Patrick Rothfuss, with his *The Kingkiller Chronicle*, carries forward Tolkien's legacy with a sophisticated level of world-building and lyrical storytelling, akin to the mythic histories found in *The Silmarillion* (Flieger, 2002). These American fantasy authors have embraced Tolkien's blueprint and adapted it for new generations, incorporating modern sensibilities while preserving the grandeur and depth that characterize his works.

Tolkien's Influence on Cinema and Popular Culture

Tolkien's influence, however, extends well beyond literature into cinema and popular culture. Peter Jackson's *The Lord of the Rings* and *The Hobbit* trilogies brought Tolkien's expansive world to an even broader audience, inspiring new waves of fantasy films and television series. American productions such as *The Chronicles of Narnia* (adapted from the works of Tolkien's contemporary and friend, C.S. Lewis) and HBO's *Game of Thrones* clearly reflect Tolkien's narrative techniques, focusing on epic conflicts, richly drawn characters, and moral complexity (Shippey, 2005). Even *Harry Potter*, though a very different type of fantasy, carries Tolkien's legacy, particularly in its magical worlds, mysterious artifacts, and the overarching theme of the battle between good and evil (Flieger, 2002).

The Mythological Foundations of Tolkien's Works

Tolkien's inspirations are as fascinating as his creations. He was a philologist and a professor of Anglo-Saxon at Oxford, and his deep academic background in languages and mythology significantly influenced his writing. His expertise in Old English, Norse mythology, and early European folklore directly shaped the world of Middle-earth (Shippey, 2000). The character of Gandalf, for example, was inspired by Odin, the Norse god, while the Elves were influenced by both Norse and Celtic traditions (Flieger, 2002). Moreover, Tolkien integrated elements of Christian mythology, particularly exploring themes of sacrifice, redemption, and the eternal struggle against evil (Shippey, 2000). His famous essay, "On Fairy-Stories," articulates his belief in the power of myth and storytelling as essential tools for understanding human experience, emphasizing his intention to create a "mythology for England" (Tolkien, 1964). This unique blend of inspirations not only shaped his fictional worlds but also forged a path for future fantasy authors to build upon.

Critical Essays on Tolkien's Influence

For those who seek a deeper understanding of Tolkien's mythological foundations and literary influence, several critical essays offer invaluable insights. Tom Shippey's *J.R.R. Tolkien: Author of the Century* provides an in-depth analysis of Tolkien's literary methods and mythic inspirations, highlighting the complex narrative techniques he employed (Shippey, 2000). Verlyn Flieger's *Splintered Light: Logos and Language in Tolkien's World* delves into Tolkien's use of language and the interplay of light and darkness in his mythology (Flieger, 2002). Additionally, Shippey's *The Road to*

Middle-earth is an essential resource for understanding the academic and folkloric roots of Tolkien's works, connecting his inspirations to his writing and examining the profound influence of his myth-making on the broader fantasy genre (Shippey, 2005). These texts serve as key references to explore Tolkien's mythology in depth and its lasting impact on modern storytelling.

As an admirer of Tolkien's work, I am struck by the depth of his legacy—not only in how he transformed the fantasy genre but also in how he provided a framework for understanding myth and its power to shape our perception of the world. His works continue to inspire countless creators, challenging us to see beyond the surface of storytelling to the cultural and philosophical roots that lie beneath. Through Tolkien, we are invited to embark on journeys that are not only adventurous but deeply resonant, reflecting the timeless truths of human experience.

References

Flieger, V. (2002). *Splintered Light: Logos and Language in Tolkien's World*. Kent State University Press.
Shippey, T. (2000). *J.R.R. Tolkien: Author of the Century*. HarperCollins.
Shippey, T. (2005). *The Road to Middle-earth*. Houghton Mifflin.
Tolkien, J.R.R. (1964). *Tree and Leaf*. HarperCollins.

This response is so much better. As we instructed, ChatGPT simply used the example we gave it as inspiration for reworking the introduction it wrote already. Essentially, it's been taught to learn the same way we do. We learn from greater examples and take their influences with us into our own work, not unlike how other authors learned from Tolkien or Shakespeare.

Conclusion

In this chapter, we looked at how ChatGPT can be trained to prepare an academic student for intensive writing assignments like writing a thesis. In the next chapter, we'll dive into how ChatGPT can help with data analysis, allowing you to make informed decisions based on the results. We'll look at practical examples such as explaining the contents of a Privacy Policy PDF file, identifying and finding information from a photo, and creating a flashcard set from an article in another language.

CHAPTER 6

Prompts to Make Your Life Easier with Data Analysis

The words "data analysis" may sound like a very technical term, but in actuality, analyzing data is something we all do on a daily basis. For example, we analyze data when we see it's supposed to rain outside and therefore choose to bring an umbrella to work. We're analyzing data when we read an email so that we can formulate a response. We do it all the time without thinking.

So, when we're talking about using ChatGPT for data analysis, what we're really talking about is giving ChatGPT information to understand **first**, then proceed to ask it to do something with the information provided. This is useful for various purposes, such as understanding context in a conversation and summarizing the important details, reading the terms and conditions of an application and explaining what they mean in simpler terms, or looking at a photo and identifying what's in it. Not only that, but we can present this information in different formats. ChatGPT can read Word documents, PDF files, and even Excel files. So let's take a look at what we can do with ChatGPT once we've given it some data to analyze.

© Lydia Evelyn 2025
L. Evelyn, *Making ChatGPT Work for You*, https://doi.org/10.1007/979-8-8688-1445-7_6

What will be covered in this chapter:

- Using ChatGPT to summarize the contents of a conversation in a Slack channel and give you the most important details

- Giving ChatGPT a PDF of Facebook's privacy policy to find out what you're really being asked to agree to

- Identifying a photo of a bug from the garden. Is it dangerous or delightful? Maybe a little of both?

- Using ChatGPT to create a flashcard set of vocabulary words from an article in a foreign language

Staying in the Loop in a Slack Channel

Most workplaces use some form of instant messaging, be it Slack, Microsoft Teams, or even Discord. It's useful for keeping up with company news, collaborating with team members, and holding friendly conversations. But the fate of all group messaging channels is to become cluttered with messages quickly, especially when something important happens—be it good or bad. For example, it could be the fact that the sales team just landed a huge deal with an important customer, and everyone is praising the team for their hard work. Or, there's been a security breach, and the entire company needs to respond to the crisis.

Or—and we've all been there—you were on vacation and dared not look at a single Slack message while you were offline. You know that when you get back, you have a lot of scrolling to do to get caught up on what everyone's talking about. Thankfully, ChatGPT can easily summarize a conversation for you and highlight the most important details so you don't have to feel left out.

Let's look at an example Slack conversation to test this out with, as shown in Listing 6-1.

Listing 6-1. A somewhat lengthy Slack channel conversation about weekend plans

#marketing-chit-chat

Emily:
Happy Friday, everyone! Any fun plans for the weekend?

Mike:
Hey Emily! I'm planning to hit the beach if the weather stays nice. How about you?

Emily:
Sounds awesome! I'm thinking of trying out that new sushi place downtown. Anyone wants to join?

Lisa:
I'm in! I've been craving sushi all week. What time are you thinking?

Emily:
How about 7 PM tomorrow?

James:
Count me in too! I'll bring my expertise in tasting sushi

David:
Haha, James, I hope your sushi expertise is better than your karaoke skills from last weekend!

James:
Touché, David. I promise no singing this time, just eating!

Sarah:

I wish I could join, but I have a family thing tomorrow. Next time for sure!

Emily:

We'll miss you, Sarah! We'll have some sushi in your honor. 😊

Mike:

On a different note, anyone up for a hike on Sunday morning? I found this great trail about an hour away.

Lisa:

That sounds fun, Mike! I'm up for it. Need to burn off all the sushi calories anyway. 😁

David:

I'll join too. What time are we thinking?

Mike:

Let's meet at 8 AM at the trailhead. Early start, but totally worth it.

James:

8 AM on a Sunday? Mike, you're killing me. But fine, I'm in. I'll bring coffee for everyone.

Emily:

Looks like a productive weekend for the team! Can't wait for tomorrow and Sunday.

Lisa:

By the way, did anyone see the new episode of that marketing show on Netflix? So many good ideas!

James:

Yes! I binged the whole series. I think we could try out some of those campaign strategies.

Sarah:

I haven't seen it yet. Now I know what I'm doing on Sunday evening. Thanks for the recommendation!

Emily:

Alright, see you all tomorrow at 7 PM for sushi, and those up for the hike, see you Sunday morning!

David:

Looking forward to it! Have a great weekend, everyone!

Mike:

Don't forget to bring your hiking boots and your best sushi appetite. Have a good one!

As you can see, Listing 6-1 shows a casual conversation between colleagues. If you walked away from the desk for a few moments and saw this conversation ongoing and wanted to add something, scrolling all the way to the top while messages are still going back and forth might be quite annoying. Let's see how ChatGPT can make this process easier.

Asking ChatGPT to Summarize a Conversation from a Chat Window

We're going to ask ChatGPT to condense what's being said and make it palatable for someone just trying to catch up. Take a look at how we did this in Prompt 6-1.

Note   For the sake of brevity, we indicate where we've given ChatGPT the content of the Slack conversation shown in Listing 6-1 with "[Slack conversation]" in Prompt 6-1.

Prompt 6-1. Asking ChatGPT to summarize a conversation too long to read

I was away from my desk for an hour on lunch break and I missed a lot of conversation in the #marking-chit-chat channel. Here's the conversation that was had:

###

[Slack conversation]

###

Briefly summarize the most important parts of the conversation so I can catch up.

Now let's look at how ChatGPT interprets the important parts of the conversation from Listing 6-1, as seen in Response 6-1.

Response 6-1. ChatGPT gives a brief, easy-to-read summary of a long conversation

Here's a summary of the conversation in the #marketing-chit-chat channel:

- **Weekend Plans:**
  - **Emily:** Plans to try a new sushi place downtown and invited others to join at 7 PM tomorrow.
  - **Mike:** Plans to hit the beach and suggested a hike on Sunday morning at 8 AM.
- **Responses:**
  - **Sushi Outing:**
    - **Lisa, James, and David** are joining Emily.
    - **Sarah** can't make it due to a family commitment.
  - **Hike:**

- **Lisa, David, and James** are joining Mike. James will bring coffee for everyone.

- **Other Topics:**
  - **Lisa** mentioned a new marketing show on Netflix with good ideas.
  - **James** has watched it and suggested implementing some strategies.
  - **Sarah** plans to watch the show on Sunday evening.

The team is looking forward to a productive and fun weekend with both the sushi outing and the hike planned.

As you can see, ChatGPT was able to succinctly summarize the conversation, making it easy to catch up with anything we might have otherwise missed.

Now, we've seen how ChatGPT can explain plain text, but it can also read and explain text from a file! Let's look at how that works by reviewing the contents of a PDF containing the terms of a privacy policy.

Understanding Terms and Conditions Before You Agree to Them

When browsing the Internet, there are always pop-ups that ask for consent before you continue using the site. If you didn't know, if you don't click "disagree," your continued use of the platform is a form of agreement. But if you really knew what you were agreeing to, you might think twice before using certain sites. Unfortunately, sites with lengthy terms and conditions sometimes rely on the fact that the legal jargon is too complex to understand and expect people to agree without much thought, which is what happens in most cases.

To protect yourself and your data on the Internet, it would be best to know what you're signing up for. This is particularly true for content creators that may not fully understand their rights to their content once it's posted using different social media services.

So let's take a look at what Facebook includes in their terms and services agreements and see what they're really asking you to accept. In this example, we're actually going to be using a functionality we touched on in Chapter 4 when we uploaded a PDF to ChatGPT to understand the style and tone of an academic thesis. This time, we're going to see in more detail how ChatGPT can be used to read documents like a PDF and tell you about its contents.

Looking at Facebook's Privacy Policies

ChatGPT can read and understand the contents of a variety of documents. You can upload Word documents, Excel files, PDFs, CSVs, and more. In this case, we're going to give ChatGPT a PDF containing the privacy policies for all Meta applications. In case you didn't know, Meta is the parent company of Facebook, Instagram, and WhatsApp, so since they have billions of users across all of these platforms, you've mostly likely already agreed to this policy. So, let's find out what Meta is really asking us to agree to when we use one of their apps.

First, take a look at Figure 6-1 to see what it looks like when we ask ChatGPT to read a file. Figure 6-1 shows our conversation with ChatGPT after uploading our PDF.

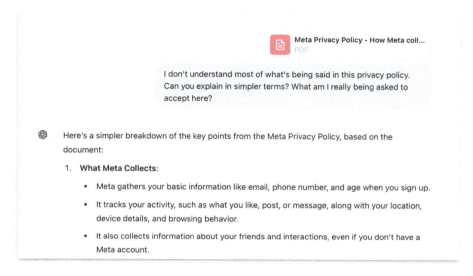

Figure 6-1. *Asking ChatGPT to read a PDF file*

As you can see from Figure 6-1, uploading a PDF to ChatGPT shows up as another text bubble above our prompt. ChatGPT then reads the document and responds to our question. It's that simple! The text in Figure 6-1 can be seen in Prompt 6-2.

Prompt 6-2. Asking ChatGPT to explain a privacy policy

```
I don't understand most of what's being said in this privacy
policy. Can you explain in simpler terms? What am I really
being asked to accept here?
```

Before we get into the response from ChatGPT, we're going to look at a snippet of the Facebook privacy policies, as shown in Listing 6-2.

Listing 6-2. A snippet of Facebook's privacy policy

Highlights
This Privacy Policy explains how we collect, use and share
your information. It also describes how long we keep your
information and how we keep it safe when we
transfer it.

- The Privacy Policy also lets you know your rights
- This Privacy Policy applies to what we call Meta Products,
  or Products. Our Products include Facebook, Instagram, and
  Messenger.
- It also applies to Meta Quest if you log in with a
  Meta account

We at Meta want you to understand what information we collect,
and how we use and share it. That's why we encourage you to
read our Privacy Policy. This helps you use
Meta Products in the way that's right for you.

In the Privacy Policy, we explain how we collect, use, share,
retain and transfer information. We also let you know your
rights. Each section of this Policy includes helpful examples
and simpler language to make our practices easier to
understand. We've also added links to resources where you can
learn more about the privacy topics that interest you.
It's important to us that you know how to control your privacy,
so we also show you where you can manage your information in
the settings of the Meta Products you use.
You can update these settings to shape your experience.

Now let's look at the explanation ChatGPT gives us on what the privacy
policy is really saying, as shown in Response 6-2.

Response 6-2. ChatGPT explains in simpler terms what's in Facebook's privacy policy

Here's a simpler breakdown of the key points from the Meta Privacy Policy, based on the document:

1. What Meta Collects:
- Meta gathers your basic information like email, phone number, and age when you sign up.
- It tracks your activity, such as what you like, post, or message, along with your location, device details, and browsing behavior.
- It also collects information about your friends and interactions, even if you don't have a Meta account.

2. Why They Collect Information:
- To personalize your experience, such as showing relevant posts, ads, or recommendations.
- To improve their products and fix issues.
- For safety, like detecting fraud or harmful behavior.
- To provide insights to advertisers and partners on how effective their ads are.

3. How They Use Your Information:
- Meta uses your data for personalization, advertising, and content recommendations.
- It shares this information with business partners, advertisers, and integrated third-party services you choose to interact with.
- Some data is used for research and improving AI technologies.

4. Your Choices:
- You can adjust settings to limit what data Meta collects or how it's used (e.g., ad preferences or location sharing).
- However, refusing some data collection might affect how the platform works for you.

5. Sharing with Others:
- Information you share can be seen by your selected audience but may also be reshared by others.
- Public information like your username or comments on public pages can be accessed by anyone, even off Meta platforms.

6. Third-Party Interactions:
- If you use third-party apps or log in with Facebook or Instagram, those apps might access some of your data.
- Meta partners may share your data with others based on their own privacy policies.

7. Retention:
- Meta keeps your information as long as necessary for the purposes outlined, or as required by law.

In Summary
By agreeing to this policy, you're allowing Meta to collect, use, and share your data to personalize your experience, show targeted ads, and enhance their services. Some of your data will also be shared with advertisers and other businesses. You have some control over your settings, but refusing certain data uses might limit features or personalization.

In the simpler version of the privacy policy, we have a much stronger idea of what kind of power we have over our data when we use Facebook if we take the steps necessary to exert that right. We also get to see plainly what Facebook is allowed to do with our data and what kinds of third parties are being allowed to view the data as well.

Understanding how your data is being used by other companies is an incredible way to improve your online safety, and it's an excellent idea to use ChatGPT to educate yourself.

But by now we know that ChatGPT isn't just able to examine text files, it can understand information from an image as well. So let's ask ChatGPT to help us figure out what bug is in a photo from a garden and if it's a threat to our begonias!

Asking ChatGPT to Explain What's in a Photo

With the ability to send ChatGPT photos, you can ask it for more information on what's being featured in the image. Let's see what that might look like.

What Bug Is in This Photo?

As a gardener finding odd insects and creatures in their garden, you might be concerned about seeing a bug you don't recognize on your plant leaves. Let's look at how we can use ChatGPT to find out what might be a threat to our garden and what steps we should take to proceed.

This time, we're going to look at a back-and-forth interaction with ChatGPT. Figure 6-2 is the image we used in our prompt. Listing 6-3 is the conversation we had with ChatGPT.

Figure 6-2. *An image of some kind of bug I found in my garden*

That's a pretty big beetle! If it was any larger, I'd ask it to pay rent. I sure hope the garden is safe. Let's ask ChatGPT. Listing 6-3 is the entire conversation we had with ChatGPT to find out. Note that we've already uploaded the photo to ChatGPT in the first prompt.

Listing 6-3. A conversation with ChatGPT identifying a beetle found in a garden

User:
I found this bug in my garden. What is it?
ChatGPT:
The beetle in your image appears to be a European rhinoceros beetle (Oryctes nasicornis). These beetles are characterized

by their large size and distinctive horn on the males. They are generally harmless and are found in gardens, feeding on decaying wood and other organic matter.

If you have any specific concerns or need more information about the beetle, feel free to ask!

User:

Are they a threat to my garden?

ChatGPT:

European rhinoceros beetles are generally not a threat to your garden. Their larvae feed on decaying wood, compost, and other organic matter, which can actually help in breaking down and recycling nutrients in the soil. Adult beetles typically do not feed or cause damage to plants.

However, if you have a lot of decaying wood or organic matter in your garden, you might see an increase in their population. Overall, they are considered beneficial insects due to their role in decomposition and nutrient cycling.

This is pretty impressive! Not only can we rest easy knowing our begonias are safe, we know that we're doing something right in our backyard's ecosystem.

But ChatGPT can do more than just identify an object in a photo. Since ChatGPT can search the web, it can also *find* items from a picture. Let's test this out by asking ChatGPT to help us shop for an outfit based upon an image from Pinterest.

Where Can I Buy the Outfit in This Photo?

People that browse the Internet and find outfits they like often struggle trying to recreate the look themselves. It's a better idea to use ChatGPT to find out where the clothes came from. Or, you might be able to find pieces similar. For this, ChatGPT will be using the web search feature, which is an impressive and useful feature for this purpose. For this example, we're going to break up our prompt into two steps. This allows ChatGPT to execute one step at a time rather than completing both tasks at once. Basically, ChatGPT might get confused otherwise. That being said, take a look at Prompt 6-3.

Prompt 6-3. Looking to recreate a look from a photo

```
Help me recreate the outfit from this photo. Describe the
individual pieces in this outfit for me!
```

Figure 6-3 is the image we used for our prompt.

Figure 6-3. *A picture of an outfit we'd like to buy for ourselvesCredit: Freepik*

156

Asking ChatGPT to describe the outfit first allows it to actually define what we're looking for, which sets us up for our next prompt, as you'll see below:

Now give me links to websites selling items just like the ones you've described.

Now that we've defined what we're looking for, let's see what ChatGPT suggests so that we can put the outfit together ourselves (Response 6-3).

Response 6-3. ChatGPT provides a list of sources to shopping websites

To help you recreate the outfit, here are some items that closely match the described pieces:

1. White Fitted Tank Top:
   - Active Fitted Tank Top - White by Cotton On: This snug-fitting tank top with thin straps is made from a lightweight, stretchy fabric, ideal for layering.

2. High-Waisted Camel Shorts:
   - Camel Woven High Waisted Tailored Shorts by PrettyLittleThing: These tailored shorts in a warm camel tone feature a high-waisted design with a fabric belt and a circular buckle, offering a chic and structured look.

3. White Button-Down Shirt (Worn Open):
   - Classic Cotton Button-Down Shirt in White by DAILYLOOK: This lightweight, slightly oversized white shirt is perfect for layering and can be worn unbuttoned over the tank top.

4. Gold Layered Necklaces:
   - Layered Necklace Set - Delicate Gold Layer Necklace Set of 3 by Etsy: This set includes multiple delicate gold necklaces of varying lengths, each with small pendants, allowing you to achieve a layered look effortlessly.

5. Gold Hoop Earrings:
   - Chunky Gold Hoops - Twisted Gold Hoops Thick Hoop Earrings by Etsy: These medium-sized thin gold hoop earrings add a classic touch to any outfit.

6. Beige or Transparent Sunglasses:
   - Sunglasses Ralph by Ralph Lauren RA5298U: These stylish, oversized sunglasses feature a slightly transparent beige frame, adding a trendy accessory to your ensemble.

7. Small Brown Leather Handbag:
   - Brown Leather Purse Crossbody Small: This structured caramel brown handbag with gold hardware is both functional and stylish, complementing the outfit perfectly.

8. Red String Bracelet:
   - Red String Bracelet by Etsy: This simple red thread bracelet adds a subtle touch of color and is adjustable for a comfortable fit.

Please note that product availability may vary, and it's advisable to check the current stock and shipping options to your location.

Sources:
https://www.macys.com/shop/womens-clothing/all-womens-clothing/womens-tops/Top_style/Tank Tops?id=255
https://www.ocsocialbutterfly.com/collections/all

```
https://thechicsavvy.com/edyta-siedleckas-trendsetting-looks/
https://www.forever21.com/us/shop/catalog/category/f21/promo_ma
ster?page=380&srule=most-popular
https://www.fashiongo.net/
https://punchyvaquera.com/collections/all/gold-hoops?srsltid=Af
mBOor37JkhXuEOF6EhKstr1jaqWcE8GRUVwbUOF74hRu9I4a7sHLo7
https://www.kellyfields.com/collections/all/enewton-style
https://www.distinctivelyhers.com/collections/all?srsltid=AfmBO
orfzoMO8JyjmcCARLtzZrEJZK94DtdlaToLjOwcyw-1WOuF-UE7
https://www.buzzfeed.com/lily_burrill/fashion-vibe-looks-
expensive-but-really-isnt
```

ChatGPT's suggestions here are great! Now let's try something different. ChatGPT is capable of understanding many languages, so why not put this capability to the test?

In the next example, we're going to use ChatGPT to create a flashcard set from an article in French.

Using ChatGPT to Learn a Language by Creating a Flashcard Set

Let me tell you something about myself. I'm a language buff. I speak English, Spanish, and Catalan fluently, as well as French at an intermediate level. Speaking from experience, when you're learning a language, one of the best strategies to learning quickly is to start immersing yourself in the media of your target language. You can do this by watching films and shows or by reading various forms of simple literature, like children's books or short articles. This is because immersion shows you how the language is used on a day-to-day basis.

If you wanted to take this learning strategy a step further, you could make note of words or phrases you don't understand and create a flashcard deck to practice with. Apps like Quizlet or Anki use scientifically proven methods to help you practice and retain the information you're trying to memorize.

As someone that's currently studying Japanese, I can personally testify to the efficacy of this method.

Before now, I've always had to create my flashcard sets myself. But ChatGPT now makes this process so much easier. By giving it an article written in a language you're interested in learning, you can have it create a vocabulary list of useful words and phrases from the article that you can then put into your flashcard deck.

Sounds interesting, right? Let's dive right in.

Using ChatGPT to Get Vocabulary from an Article in French

We can have ChatGPT read an entire article just by providing the link! So that's precisely what we're going to instruct ChatGPT to do here. After all, it would be quite cumbersome to have to copy and paste the entire article ourselves! We include a link, as you can see in Prompt 6-4, which leads to an article from *Le Monde* (the largest newspaper in France) by the author Par Clément Ghys.

Prompt 6-4. Asking ChatGPT to create a French flashcard deck

```
I am trying to learn French and I need you to suggest
vocabulary to write in flashcards. I have an article I'm
interested in reading, but I don't understand anything. I want
you to create a list of vocabulary words with translations
of the words most relevant to the topic of the article, the
most commonly used phrases with grammatical explanations and
```

a brief summary of what the article was about so I can title my deck of flashcards accordingly. I want a nice long list of relevant nouns, verbs and grammatical notes that I can add to my flashcards.

I'm not fluent so be sure to provide your tips and advice in English!

https://www.lemonde.fr/m-styles/article/2024/07/21/l-hotel-del-coronado-decor-d-un-vaudeville-culte_6254833_4497319.html

Now that we've specified what we're looking for in a good flashcard set, let's see how ChatGPT complies. Response 6-4 includes a vocabulary list based upon content from the article we provided.

Response 6-4. ChatGPT creates a vocabulary list with grammatical tips

The article from Le Monde discusses the Hotel del Coronado, affectionately known as "The Del," located near San Diego, California. In 1959, director Billy Wilder chose this iconic wooden structure, built in 1888, as the filming location for his comedy "Some Like It Hot." The hotel, with its distinctive turrets and canopies, provided a picturesque backdrop, standing in for a Florida setting in the film. Over the years, "The Del" has been a favored retreat for numerous Hollywood stars, including Douglas Fairbanks, Charlie Chaplin, and Marilyn Monroe. Today, it remains a popular destination for tourists, who are drawn to its rich history and scenic beachfront location.

To assist you in understanding and engaging with the article, here's a curated list of relevant French vocabulary, including nouns, verbs, and phrases, along with their English translations and grammatical notes:

Nouns (Les Noms):

- **l'hôtel (masc.)** – hotel
- **le décor** – set, scenery
- **le vaudeville** – vaudeville (a genre of comedy)
- **la comédie** – comedy
- **le réalisateur** – director
- **le tournage** – filming
- **la production** – production
- **la plage** – beach
- **la mafia** – mafia
- **le musicien / la musicienne** – musician
- **le milliardaire** – billionaire
- **le déguisement** – disguise
- **la course-poursuite** – chase
- **le restaurant** – restaurant
- **le symposium** – symposium
- **le mafieux** – mobster
- **la conférence** – conference
- **l'opéra (masc.)** – opera
- **la star** – star (celebrity)
- **le salon** – lounge
- **le tournage** – filming
- **la presse à scandales** – tabloid press
- **le ragot** – gossip
- **le dessert** – dessert
- **le soufflé** – soufflé
- **la vanille** – vanilla
- **la meringue** – meringue
- **le touriste** – tourist
- **la chambre** – room
- **l'histoire (fem.)** – history

162

- **la plage** – beach
- **la chambre** – room
- **le fantôme** – ghost

Verbs (Les Verbes):

- **tourner** – to film
- **imaginer** – to imagine
- **juger** – to judge
- **fuir** – to flee
- **se travestir** – to disguise oneself
- **rejoindre** – to join
- **rencontrer** – to meet
- **tomber amoureux** – to fall in love
- **décider** – to decide
- **épouser** – to marry
- **s'éprendre** – to fall in love
- **se cacher** – to hide
- **se déguiser** – to disguise oneself
- **se piquer** – to claim
- **séduire** – to seduce
- **se faire passer pour** – to pretend to be
- **claquer** – to slam
- **accueillir** – to welcome
- **maquiller** – to disguise
- **sortir** – to release (a film)
- **voir** – to see
- **venir** – to come
- **s'amuser** – to have fun
- **devoir** – to have to
- **refaire** – to redo
- **imposer** – to impose
- **servir** – to serve

- **attirer** - to attract
- **séduire** - to seduce
- **revenir** - to come back
- **aller** - to go
- **se rendre** - to go
- **voir** - to see
- **être réputé** - to be reputed

Common Phrases with Grammatical Explanations:

- **"Certains l'aiment chaud"** - "Some Like It Hot." *Note:* In French, movie titles are often translated differently from their literal meanings.
- **"a jugé plus pratique de ne pas quitter la Californie"** - "deemed it more practical not to leave California." *Note:* "juger" means "to judge" or "to deem," and is followed by an infinitive verb.
- **"deux musiciens recherchés par la mafia"** - "two musicians sought by the mafia." **Note:** "recherchés" is the past participle of "rechercher," used here as an adjective agreeing in gender and number with "musiciens."
- **"en se travestissant et en rejoignant un orchestre de femmes"** - "by disguising themselves and joining a women's orchestra." *Note:* The gerund form "en + present participle" indicates simultaneous actions.
- **"dont l'un tombe amoureux"** - "one of whom falls in love." *Note:* "dont" is a relative pronoun indicating possession or association.
- **"a décidé qu'une fois à Miami elle rencontrerait un homme riche et l'épouserait"** - "decided that once in Miami she would meet a rich man and marry him." *Note:* The conditional tense "rencontrerait" and "épouserait" expresses a future action viewed from a past perspective.

- **"sera le théâtre d'un merveilleux vaudeville"** - "will be the scene of a marvelous vaudeville." *Note:* "être le théâtre de" is an idiomatic expression meaning "to be the scene of."
- **"ne comprenant pas qu'un homme se cache sous ce déguisement"** - "not realizing that a man is hiding under this disguise." *Note:* "ne... pas" forms the negation, and "se cacher" is a reflexive verb meaning "to hide oneself."
- **"se pique, pour séduire la musicienne, de se ... "** - "pretends, to seduce the musician, to pass himself off as a depressed and impotent billionaire." *Note:* "se piquer de" means "to claim to," and "se faire passer pour" means "to pretend to be."
- **"Les portes claquent, les ... "** - "Doors slam, hallways are the setting for chases." *Note:* "claquer" means "to slam," and "course-poursuite" is a compound noun meaning "chase."
- **"le restaurant accueille un symposium de mafieux déguisé en conférence des Amis de l'Opéra italien"** - "The restaurant hosts a symposium of mobsters disguised as a conference of the Friends of Italian Opera."
  Note: "accueillir" means "to host" or "to welcome," and "déguisé" is a past participle used as an adjective to describe "symposium."

- **"Elle décide de séduire un milliardaire"** - "She decides to seduce a billionaire."
  Note: The infinitive "séduire" follows the verb "décider de," which introduces an action decided upon.

- **"La star américaine passe dans les salons de l'hôtel"** - "The American star passes through the hotel lounges."
  Note: "passer dans" means "to pass through" or "to visit a place."

- **"Les touristes viennent pour découvrir son histoire"** – "Tourists come to discover its history."
Note: "venir" is often used with an infinitive verb to indicate purpose.

Additional Grammatical Notes:

- **Gender Agreement:** In French, nouns and adjectives must agree in gender and number with the subject or object they describe. For example, **"un milliardaire"** (a male billionaire) becomes **"une milliardaire"** (a female billionaire).

- **The Conditional Tense:** Used to express what would happen under certain circumstances. For example:
  - **"Elle rencontrerait un homme riche."** – "She would meet a rich man."

- **Reflexive Verbs:** Verbs like **se cacher** (to hide oneself) and **se déguiser** (to disguise oneself) require reflexive pronouns, which change based on the subject (e.g., je me, tu te, il/ elle se).

- **The Gerund Form:** Created with **en** + present participle (e.g., **en se travestissant** – "by disguising oneself") and indicates simultaneous actions.

- **Relative Pronouns:**
  - **"dont"** – used for possession or association ("of whom," "of which").
  - **"que"** – used for the object of a verb ("that," "which").
  - **"qui"** – used for the subject of a verb ("who," "which").

- **Negation:** Formed with **ne... pas** (e.g., **"Il ne comprend pas."** – "He does not understand").

Summary for Your Flashcards

This article focuses on the **Hotel del Coronado** as a significant filming location for the classic movie **"Some Like It Hot."** It describes the hotel's unique architecture, its historical significance, and its appeal as a destination for both tourists and Hollywood stars. The article also delves into the plot of the film and how the hotel served as a backdrop for the comedic and romantic escapades of the characters.

Suggested Title for Your Flashcards

"L'Hôtel del Coronado et 'Certains l'aiment chaud'"

You can now create your flashcards with these vocabulary words, grammatical notes, and phrases to enhance your understanding of the article while practicing French! Let me know if you'd like additional help with any part of this. 😊

This is incredibly helpful! ChatGPT provides us an impressive list of vocabulary for a flashcard set along with grammatical notes we can study on our own. If I may say, the response we got from ChatGPT was *incroyable*.

Conclusion

In this chapter, we got a better understanding of how data analysis can be used for summarizing text, understanding complex language in a document, explaining what's in a photo, and even creating a flashcard set from an article.

In previous chapters, we've discussed how ChatGPT can be taught to perform tasks. However, in the next chapter, we're going to look at how ChatGPT can also be used as a learning tool to teach *you* new skills by providing you with learning material and helpful criticism.

CHAPTER 7

Prompt ChatGPT to Learn New Skills

In previous chapters, we've looked at how we could teach ChatGPT new information to make it more useful. However, in this chapter, we're going to see how ChatGPT can teach *us* something new.

For example, in the last chapter, we asked ChatGPT to read over the contents of a privacy policy and explain them in simpler terms. But we didn't really learn anything new about privacy policies as a whole to better inform our decisions on agreeing to such terms in the future.

Typically, when we don't know the answer to a question, our first instinct is to consult Google for an answer. But the issue with a Google search result is that you're limited to what is offered by any particular web page, image, or video. You can't ask questions, get clarification, or get more information other than what's on the page.

With ChatGPT, we can go far deeper than simply discover new information. It can be used to learn entirely new concepts. Whereas in the last chapter we had ChatGPT explain a privacy policy in simpler terms, in this chapter, we're going to have ChatGPT answer questions like how can we fulfill our obligations in a lease agreement? What exactly does copyright law do, and who does it protect? We're going to ask ChatGPT to help us create a learning plan to start from absolute zero and end up knowing how to create our own website. And we're going to continue exploring language learning capabilities within ChatGPT to turn mistakes into custom structured lessons.

© Lydia Evelyn 2025
L. Evelyn, *Making ChatGPT Work for You*, https://doi.org/10.1007/979-8-8688-1445-7_7

Let's get started.

What will be covered in this chapter:

- Using ChatGPT to understand rights and obligations within a lease agreement

- Using ChatGPT to understand copyright law so that you're not infringing on someone else's rights by accident

- Asking ChatGPT to teach you essential programming skills so you can code your own website

- Taking advantage of ChatGPT's multilingual capabilities to correct our French and turn mistakes into lessons

Using ChatGPT to Understand Important Legal Documents

In Chapter 6, we looked at how ChatGPT could be used to understand Facebook's privacy policies. What we're going to cover here is slightly different. Misunderstanding how your data is being used by a social media platform can be dangerous, but not nearly as dangerous as misunderstanding a legal document.

Understanding things such as a contract, government paperwork, or copyright law is crucial to ensuring you understand both your rights and the rights of others. Unfortunately, understanding these documents can be difficult for some people. For this reason, we're going to look at how ChatGPT can help us better understand the significance of legal documents that should be handled with care.

Getting Clarification on the Terms of a Lease Agreement Before You Sign

A lease agreement is one of the most common contracts people come across as adults. They're also easy to misunderstand, which can lead to messy contract terminations. This is especially true when the contract isn't reviewed by a lawyer, which is more and more common in the modern era.

So let's use ChatGPT to tell us more about the terms of the lease contract abbreviated in Listing 7-1. For this example, we've taken a sample contract and given it to ChatGPT in the form of a PDF.

Listing 7-1. Asking ChatGPT to help us understand the terms of a lease agreement (abbreviated)

```
PREAMBLE:
This lease contains the agreements between You and Owner
concerning Your rights and obligations and the rights and
obligations of Owner. You and Owner have other rights and
obligations which are set forth in government laws and
regulations. You should read this Lease and all of its attached
parts carefully. If you have any questions, or if you do not
understand any words or statements, get clarification. Once you
and Owner sign this Lease You and Owner will be presumed to
have read it and understood it. You and Owner admit that all
agreements between You and Owner have been written into this
Lease. You understand that any
agreements made before or after this Lease was signed and not
written into it will not be enforceable.
...
```

Interestingly, the preamble shown in Listing 7-1 suggests seeking clarification if the potential tenant doesn't understand the terms. It also says that by agreeing, reading and understanding is *assumed*. This is one of the reasons why getting a clear interpretation of a contract is crucial. Once signed, there isn't much room for negotiation of terms that have already been agreed upon simply because you misunderstood the first time.

So, taking the advice given in the contract itself, you'll see that Figure 7-1 shows what it looked like when we uploaded a PDF of our example lease agreement to ChatGPT.

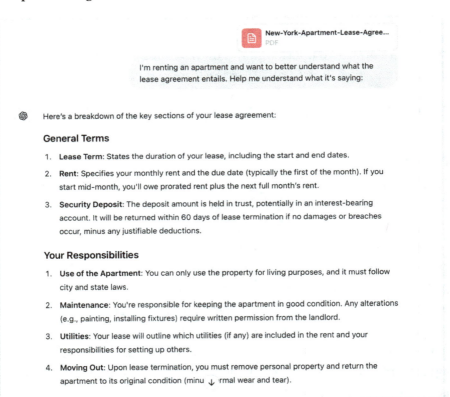

Figure 7-1. *Once again, uploading a PDF to ChatGPT*

In Prompt 17-1, we ask ChatGPT to explain the terms of the agreement.

Prompt 17-1. ChatGPT explains each section of the lease agreement

I'm planning on renting an apartment and want to better understand what the lease agreement the landlord sent me entails. Help me understand what it's saying:

Just as before, ChatGPT has analyzed the document and given us an explanation on its contents (Response 7-1).

Response 7-1. ChatGPT makes each section of the lease agreement easier to understand

Here's a breakdown of the key sections of your lease agreement:

General Terms
1. Lease Term: States the duration of your lease, including the start and end dates.
2. Rent: Specifies your monthly rent and the due date (typically the first of the month). If you start mid-month, you'll owe prorated rent plus the next full month's rent.
3. Security Deposit: The deposit amount is held in trust, potentially in an interest-bearing account. It will be returned within 60 days of lease termination if no damages or breaches occur, minus any justifiable deductions.

Your Responsibilities
1. Use of the Apartment: You can only use the property for living purposes, and it must follow city and state laws.
2. Maintenance: You're responsible for keeping the apartment in good condition. Any alterations (e.g., painting, installing fixtures) require written permission from the landlord.
3. Utilities: Your lease will outline which utilities (if any) are included in the rent and your responsibilities for setting up others.

4. Moving Out: Upon lease termination, you must remove personal property and return the apartment to its original condition (minus normal wear and tear).

Rules and Restrictions

1. Objectionable Conduct: Actions that disturb other tenants or make the building unfit for living can result in lease termination.

2. Pets: Generally prohibited unless explicitly allowed. Even with permission, pets must adhere to building-specific rules.

3. Noise and Public Areas: Excessive noise and misuse of shared spaces can lead to penalties.

4. Alterations: Major changes to the apartment (e.g., installing appliances or water-filled furniture) require the landlord's written consent.

Landlord's Rights

1. Entry: The landlord can enter the apartment with reasonable notice for inspections, repairs, or showing the property to prospective renters/buyers.

2. Default and Remedies: If you fail to pay rent or breach the agreement, the landlord can terminate the lease and seek damages.

3. Property Damage or Casualty: Rent reductions may apply if the apartment becomes unusable due to unforeseen events, but the landlord can terminate the lease in extreme cases.

Legal Notes

1. Subletting/Assigning: You generally cannot sublet or assign the lease without landlord approval.

2. Jury Waiver: Both parties waive the right to a jury trial for disputes related to the lease.

3. Insurance and Liability: The landlord is typically not liable for tenant property damage unless caused by negligence.

Let me know if you'd like help with any specific clauses or terms!

This is a great explanation! What's more, we can ask questions to better understand any given subject. Let's try asking about the tenant's responsibilities listed in the contract.

Prompt 7-2. Asking ChatGPT to clarify

I don't understand the part about the tenant's obligations on maintenance and repairs

Now let's see how ChatGPT answers our question, as shown in Response 7-2.

Response 7-2. ChatGPT helps us understand tenant responsibilities

The maintenance and repair obligations outlined in your lease can be summarized as follows:

Tenant's Obligations
1. General Care:
- You must take good care of the apartment and ensure no damage occurs beyond ordinary wear and tear.
- This includes keeping the apartment clean and avoiding any actions that might harm the property.

2. Alterations:
- You cannot make changes such as painting, wallpapering, or installing fixtures without the landlord's written consent.
- If you make any alterations (even with permission), you are responsible for restoring the apartment to its original condition before moving out. This includes removing installations like shelves, murals, or other personal modifications.

175

3. Repairs for Negligence:
- If damage occurs due to your negligence or misuse (e.g., broken appliances, plumbing issues caused by improper use), you are responsible for repair or replacement costs.

4. Moving Out:
- When leaving, you must remove all personal belongings and return the apartment in the same condition as when you moved in, except for normal wear and tear or damages from events like fire or accidents outside your control.

What Happens if Obligations Aren't Met:
- If you fail to fulfill these responsibilities, the landlord can:
  - Deduct repair costs from your security deposit.
  - Charge you for additional expenses beyond the security deposit if necessary.
  - Potentially treat the failure as a lease violation, which could have legal consequences.

This is a great example of how you can ask ChatGPT about any detail in a document and get a clear answer. Clearly, you can't ask a question so specific to your circumstance with a Google search.

As you can see, while it's always advisable to consult a lawyer for legal advice and ChatGPT cannot replace a real lawyer, it's never a bad idea to get clarity on confusing terms at any time when considering a contract or negotiation. Understanding your rights and the rights of others is important for ensuring fair treatment for everyone involved.

In fact, a law that is all too commonly infringed upon is copyright law. And unfortunately, many individuals infringe on copyright laws quite often simply due to misunderstanding how it works. So let's ask ChatGPT to help us get a clearer understanding of how to avoid making this common mistake.

Learning How to Use Music on Social Media the Right Way

In the age of social media, it's all too common for content creators to infringe on copyright laws unintentionally. One of the most common infringements is using licensed music. This happens even more now with the popularity of apps like TikTok, where music is used in almost every video.

For this reason, let's give ChatGPT an example of a copyright infringement from a YouTuber making a mukbang with music in the background to help us better understand how copyright laws work (Prompt 7-3).

Prompt 7-3. ChatGPT explains how to keep yourself from committing copyright infringement

I'm a YouTuber and I'm trying to use Espresso by Sabrina Carpenter in the background of my mukbang, but I got flagged and my video was taken down. Why did this happen?

In Response 7-3, ChatGPT offers a helpful explanation of copyright law and how YouTube enforces compliance on its platform.

Response 7-3. ChatGPT explains copyright infringement on YouTube

Your video was likely flagged and removed because "Espresso" by Sabrina Carpenter is protected by copyright. Using copyrighted music in your content without proper authorization can lead to violations of YouTube's copyright policies. YouTube employs a system called Content ID, which automatically detects and manages copyrighted material. When Content ID identifies copyrighted content in a video, the copyright owner can choose

177

to block the video, monetize it by running ads, or track its viewership statistics. In your case, the copyright owner probably opted to block the video, resulting in its removal.

To avoid such issues in the future, consider the following options:

1. Obtain Permission: Reach out to the copyright owner or their representatives to request permission to use the song in your video. This may involve licensing fees or specific usage terms.

2. Use Royalty-Free Music: Utilize music from royalty-free libraries that allow you to use tracks without infringing on copyright. Platforms like YouTube offer audio libraries with free music for creators.

3. Create Original Music: If possible, produce your own music or collaborate with musicians to create original tracks tailored for your content.

By adhering to copyright laws and YouTube's policies, you can prevent future content removals and maintain a positive standing on the platform.

This explanation is very clear and easy to understand for someone that is trying to make a successful YouTube channel, but can't seem to comprehend why their videos keep getting flagged. With ChatGPT, you can avoid accidentally losing your channel over a simple misunderstanding.

Now that we've seen how ChatGPT can help us understand legal documents, let's find out how ChatGPT can teach us essential skills for the modern technology era. Let's learn how to make a website.

Learning to Code and Create Your Own Website

Creating a website is often one of the first steps in establishing a secondary income. However, creating a website is also one of the most intimidating tasks for small business owners. Coding in general tends to be an intimidating skill for most. Online courses teaching the subject are often expensive, and encountering errors without knowing how to solve them is often the most frustrating aspect. You can see why it's one of the main reasons people give up the endeavor.

However, not only can you use ChatGPT to teach you to make your own website, but you can even send it any errors you come across and offer a solution.

Let's ask ChatGPT to take us all the way from the beginning. We're going to instruct ChatGPT to build a curriculum for our own personal online class on building a website by ourselves. Check out Prompt 7-4.

Prompt 7-4. A simple prompt to get a detailed lesson plan!

```
I want to learn how to make a website, but I've never coded
before. Create a lesson plan for me to learn the basics of
programming and web development.
```

As you can see, our prompt is very simple, but the answer we got from ChatGPT in Response 7-4 is very detailed and thorough.

Response 7-4. ChatGPT provides a thorough month-long lesson plan so you can learn to code

Lesson Plan: Introduction to Programming and Web Development

Week 1: Introduction to Programming Concepts

1. Day 1: Understanding Programming
- Learn what programming is and why it's useful.
- Introduction to programming languages.
- Resources: https://www.khanacademy.org/computing/computer-programming

2. Day 2: Getting Started with Python
- Install Python and set up the environment.
- Learn basic syntax: variables, data types, and basic operators.
- Resources: https://www.python.org/about/gettingstarted/

3. Day 3: Control Structures
- Learn about conditional statements (`if`, `else`, `elif`).
- Introduction to loops (`for`, `while`).
- Practice with simple problems.
- Resources: https://www.w3schools.com/python/python_conditions.asp

4. Day 4: Functions and Modules
- Understand functions, parameters, and return values.
- Introduction to modules and how to import them.
- Resources: https://realpython.com/defining-your-own-python-function/

5. Day 5: Practice Day
- Work on small projects or exercises to reinforce learning.
- Resources: https://www.hackerrank.com/domains/tutorials/10-days-of-python

6. Day 6: Debugging and Error Handling
   - Learn about common errors and how to debug them.
   - Introduction to try-except blocks.
   - Resources: https://www.w3schools.com/python/python_try_
     except.asp

7. Day 7: Review and Q&A
   - Review the week's topics and clarify doubts.
   - Resources: Join a coding community like https://
     stackoverflow.com/ for support.

Week 2: Introduction to Web Development
1. Day 1: Understanding the Web
   - How websites work: Frontend vs. Backend.
   - Introduction to HTML.
   - Resources: https://developer.mozilla.org/en-US/docs/Learn/
     Getting_started_with_the_web/HTML_basics

2. Day 2: HTML Basics
   - Structure of an HTML document.
   - Common HTML tags: headings, paragraphs, links, images.
   - Resources: https://www.w3schools.com/html/

3. Day 3: Introduction to CSS
   - What is CSS and how it's used.
   - Basic syntax and selectors.
   - Resources: https://www.w3schools.com/css/

4. Day 4: Styling with CSS
   - Learn about colors, fonts, and layout.
   - Introduction to the box model and positioning.
   - Resources: https://developer.mozilla.org/en-US/docs/Learn/
     CSS/CSS_layout

5. **Day 5: Introduction to JavaScript**
    - What is JavaScript and its role in web development.
    - Basic syntax, variables, and data types.
    - Resources: https://www.javascript.com/learn

6. **Day 6: JavaScript Basics Continued**
    - Control structures, functions, and events.
    - Manipulating the DOM (Document Object Model).
    - Resources: https://developer.mozilla.org/en-US/docs/Web/JavaScript/Guide

7. **Day 7: Building a Simple Website**
    - Combine HTML, CSS, and JavaScript to build a simple static website.
    - Resources: https://www.freecodecamp.org/learn/responsive-web-design/responsive-web-design-projects/build-a-tribute-page

Week 3: Advanced Web Development Concepts

1. **Day 1: Responsive Design**
    - Learn about responsive design and media queries.
    - Make your website mobile-friendly.
    - Resources: https://developer.mozilla.org/en-US/docs/Learn/CSS/CSS_layout/Responsive_Design

2. **Day 2: Introduction to Version Control**
    - Learn about Git and GitHub.
    - Basic commands: clone, commit, push, pull.
    - Resources: https://guides.github.com/activities/hello-world/

3. **Day 3: Introduction to Web Hosting**
    - Learn how to deploy your website.
    - Explore free hosting options.
    - Resources: https://pages.github.com/

4. Day 4: Introduction to Frameworks
   - Learn about CSS frameworks like Bootstrap.
   - Basic usage and components.
   - Resources: https://getbootstrap.com/docs/5.0/getting-started/introduction/

5. Day 5: Introduction to Backend Development
   - Learn about server-side programming with Python (using Flask).
   - Set up a basic web server.
   - Resources: https://flask.palletsprojects.com/en/2.0.x/tutorial/

6. Day 6: Connecting Frontend and Backend
   - How to make requests to the backend.
   - Introduction to APIs and JSON.
   - Resources: https://developer.mozilla.org/en-US/docs/Learn/JavaScript/Client-side_web_APIs/Introduction

7. Day 7: Final Project
   - Build a full-stack web application.
   - Resources: Use previous lessons and online documentation for guidance.

Continuous Learning
- **Join Coding Communities:** Participate in forums like Stack Overflow, Reddit, or coding Slack groups.
- **Build Projects:** Continuously challenge yourself with new projects.
- **Online Courses:** Enroll in courses from platforms like Coursera, Udemy, or freeCodeCamp to deepen your knowledge.
- **Read Documentation:** Regularly read documentation and articles to stay updated with new technologies and best practices.

```
By following this plan, you'll gain a solid foundation in both
programming and web development, enabling you to create and
deploy your own websites.
```

This answer from ChatGPT not only gives you a guide that will last you an entire month of learning, it also gives you various free sources to learn from to follow the lesson plan. It even gives you a guide on where to go from there for further learning.

Let's look at other ways that ChatGPT can help us learn essential skills. Let's use ChatGPT to help us learn French by improving upon our mistakes.

Using ChatGPT As a Multilingual Tutor by Creating Lessons from Your Mistakes

We've just seen that ChatGPT can create a lesson plan including helpful resources to help you follow the plan to completion. We could ask for a similar plan for learning a language by creating a similar prompt as we did for the previous examples.

However, assuming you're already on the journey to learning a language, you'll get to a point where you'll realize that a key aspect of learning is getting feedback. After all, you can learn as much vocabulary as you'd like, but if you're not using these words in sentences properly, you're not making much progress. So, let's use ChatGPT to help us with corrections when we've made a mistake.

Write a Diary Entry in Your Target Language, Then Get Correction

This is going to be a fun exercise. If you didn't know, for people learning a language, it's often a good idea to practice writing diary entries in the target language. It allows one to practice words they use often but don't practice in their typical learning sessions.

So, let's give ChatGPT a diary entry in French, giving it our best shot. Then we'll let ChatGPT correct what we've written. Take a look at Prompt 7-5.

Prompt 7-5. Asking ChatGPT to correct our French

I'm learning French and am writing diary entries in French to practice. Please correct it for me and explain the corrections in English:

###

Aujourd'hui je allé dans un reunion avec mon chef et l'équipe de marketing, ensuite allé à déjeuner avec mes amis. J'avais beaucoup de amusé et j'espere nous avons le chance de sortir plus souvent. Apres travail, je suis allé shopping pour chaussures, mais ne pouvais pas trouve ma taille dans le paire que j'ai aimé le plus!

Now let's see how ChatGPT answered in Response 7-5.

Response 7-5. ChatGPT points out the mistakes in our French

Sure! Here are the corrections and explanations for your diary entry in French:

Corrected Diary Entry:
Aujourd'hui, je suis allé à une réunion avec mon chef et l'équipe de marketing, puis je suis allé déjeuner avec mes amis. Je me suis beaucoup amusé et j'espère que nous aurons la chance de sortir plus souvent. Après le travail, je suis allé faire du shopping pour des chaussures, mais je n'ai pas pu trouver ma taille dans la paire que j'aimais le plus !

Corrections and Explanations:

1. **"Aujourd'hui je allé"** to **"Aujourd'hui, je suis allé"**:
   - **Correction:** Use "suis" for the passé composé of "aller" (je suis allé).
   - **Explanation:** "Allé" is a past participle that requires the auxiliary verb "être" in the passé composé.

2. **"dans un reunion"** to **"à une réunion"**:
   - **Correction:** Change "un" to "une" because "réunion" is feminine.
   - **Explanation:** The correct preposition is "à" (to a meeting), and "une réunion" is the correct gender.

3. **"ensuite allé à déjeuner"** to **"puis je suis allé déjeuner"**:
   - **Correction:** "ensuite" to "puis" and add "je suis" before "allé".
   - **Explanation:** "Puis" is a better transition word here, and "suis allé" is necessary for correct grammar.

4. **"J'avais beaucoup de amusé"** to **"Je me suis beaucoup amusé"**:
   - **Correction:** Use the reflexive form "me suis amusé".
   - **Explanation:** The verb "amuser" is reflexive when talking about enjoying oneself.

5. **"j'espere nous avons le chance"** to **"j'espère que nous aurons la chance"**:
   - **Correction:** Add "que" and change verb tenses to "aurons".
   - **Explanation:** "Espérer que" requires a subjunctive structure with "aurons" (we will have).

6. **"Apres travail"** to **"Après le travail"**:
   - **Correction:** Add "le" before "travail".
   - **Explanation:** The definite article "le" is needed.

7. **"je suis allé shopping"** to **"je suis allé faire du shopping"**:
   - **Correction**: Add "faire du" before "shopping".
   - **Explanation**: The correct expression is "faire du shopping".

8. **"mais ne pouvais pas trouve"** to **"mais je n'ai pas pu trouver"**:
   - **Correction**: Change to "je n'ai pas pu trouver".
   - **Explanation**: "N'ai pas pu" is the correct passé composé form of "ne pas pouvoir" (could not).

9. **"ma taille dans le paire que j'ai aimé le plus!"** to **"ma taille dans la paire que j'aimais le plus !"**:
   - **Correction**: Change "le" to "la", and "j'ai aimé" to "j'aimais".
   - **Explanation**: "Paire" is feminine, so it requires "la", and "j'aimais" (imperfect tense) is better for expressing a general preference.

These corrections will make your diary entry grammatically correct and more natural in French. Keep practicing, and you'll keep improving!

This is an amazing explanation that can help us improve our French, for sure. But we can actually go even further.

Get the Most Out of Your Mistakes by Turning Them into a Lesson

Let's continue the conversation with ChatGPT so we can make the most out of learning from our diary entry (Prompt 7-6).

Prompt 7-6. Asking ChatGPT to help us learn from our mistakes

```
Turn these corrections into a lesson for me
```

Now let's see how ChatGPT responded, as shown in Response 7-6.

Response 7-6. ChatGPT turns our mistakes into an opportunity to improve!

French Diary Entry Corrections Lesson

1. Use of Passé Composé with "Être"

Original: Aujourd'hui je allé
Corrected: Aujourd'hui, je suis allé

Rule: For verbs of motion like "aller" (to go), the auxiliary verb "être" is used in the passé composé.
- **Example:** Je suis allé (I went), Tu es arrivé (You arrived)

2. Gender Agreement

Original: dans un reunion
Corrected: à une réunion

Rule: "Réunion" is a feminine noun, so it takes the feminine article "une". The correct preposition is "à" when talking about attending a meeting.
- **Example:** à une réunion (to a meeting), dans une classe (in a class)

3. Transition Words

Original: ensuite allé à déjeuner
Corrected: puis je suis allé déjeuner

Rule: Use "puis" as a transition word, and ensure the correct use of "je suis allé" for the past action.

- **Example:** Puis je suis allé (Then I went), Ensuite, j'ai fait (Then, I did)

4. Reflexive Verbs

Original: J'avais beaucoup de amusé
Corrected: Je me suis beaucoup amusé

Rule: When talking about enjoying oneself, use the reflexive form of "amuser" in the passé composé.
- **Example:** Je me suis amusé (I had fun), Nous nous sommes bien amusés (We had a lot of fun)

5. Expression of Hope and Future Events

Original: j'espere nous avons le chance
Corrected: j'espère que nous aurons la chance

Rule: Use "j'espère que" followed by the future tense "nous aurons" to express hope about future events.
- **Example:** J'espère que nous aurons (I hope we will have), J'espère que tu viendras (I hope you will come)

6. Use of Articles

Original: Apres travail
Corrected: Après le travail

Rule: Use the definite article "le" with "travail" (work).
- **Example:** Après le travail (After work), Avant la réunion (Before the meeting)

7. Expressions with "Faire"

Original: je suis allé shopping
Corrected: je suis allé faire du shopping

Rule: Use "faire du" with "shopping" to express going shopping.
- **Example:** faire du shopping (to go shopping), faire du sport (to play sports)

8. Negative Past Actions

Original: mais ne pouvais pas trouve
Corrected: mais je n'ai pas pu trouver

Rule: Use "je n'ai pas pu" for "I could not" in the passé composé.
- **Example:** Je n'ai pas pu trouver (I could not find), Je n'ai pas pu venir (I could not come)

9. Imperfect vs. Passé Composé

Original: ma taille dans le paire que j'ai aimé le plus!
Corrected: ma taille dans la paire que j'aimais le plus !

Rule: Use the imperfect tense "j'aimais" for ongoing or repeated actions in the past, and ensure gender agreement with "la paire".
- **Example:** J'aimais (I used to like), Je préférais (I preferred)

Practice Examples

1. Using "aller" in passé composé:
   - Correct: "Hier, je suis allé au cinéma." (Yesterday, I went to the cinema.)

2. Gender agreement with nouns:
   - Correct: "J'ai une voiture." (I have a car.)

3. Transition words:
   - Correct: "Puis, nous avons mangé." (Then, we ate.)

4. Reflexive verbs:
- Correct: "Elle s'est bien amusée." (She had a lot of fun.)

5. Expression of hope:
- Correct: "J'espère que tu réussiras." (I hope you will succeed.)

6. Use of articles:
- Correct: "Après le dîner, nous avons regardé un film." (After dinner, we watched a movie.)

7. Expressions with "faire":
- Correct: "Nous allons faire du vélo." (We are going to ride bikes.)

8. Negative past actions:
- Correct: "Je n'ai pas pu venir à la fête." (I could not come to the party.)

9. Imperfect vs. passé composé:
- Correct: "Quand j'étais jeune, j'aimais jouer au football." (When I was young, I liked playing football.)

By understanding and applying these rules, you'll improve your French writing skills and gain more confidence in your use of the language. Keep practicing!

In this incredible response, we're able to really comprehend the mistakes we made before and get a deeper understanding of the French language as a whole. But let's go even further.

ChatGPT Quizzes Us on Our Custom-Made Lesson

Let's cement the concepts we've learned even more and make sure we understand. We can have ChatGPT turn our lesson into a quiz! Take a look at Prompt 7-7.

Prompt 7-7. Asking to get quizzed

Create a quiz for me, but don't give me the answers.

In Response 7-7, we can see that the quiz ChatGPT offered us tests us specifically on the areas we need to improve on.

Response 7-7. Getting a custom-made French quiz from ChatGPT

Sure! Here's a quiz based on the lesson above:

French Diary Entry Corrections Quiz

Part 1: Choose the Correct Form

1. Which is the correct past tense form for "I went"?
   - a) J'ai allé
   - b) Je suis allé
   - c) J'ai allée

2. What is the correct article for "réunion"?
   - a) un
   - b) une
   - c) des

3. Which transition word is better to use after "ensuite"?
   - a) alors
   - b) mais
   - c) puis

Part 2: Correct the Sentence

4. Correct the following sentence: "Aujourd'hui je allé à un réunion."

5. Correct the following sentence: "J'avais beaucoup de amusé."

6. Correct the following sentence: "Apres travail, je suis allé shopping."

Part 3: Fill in the Blanks

7. Complete the sentence: "J'espère que nous _____ la chance de sortir plus souvent."
   - (a) avons
   - (b) aurons
   - (c) avions

8. Complete the sentence: "Je n'ai pas pu _____ ma taille."
   - (a) trouve
   - (b) trouvé
   - (c) trouvais

9. Complete the sentence: "Après _____ travail, je suis allé faire du shopping."
   - (a) le
   - (b) la
   - (c) les

Part 4: Imperfect vs. Passé Composé

10. Which form is correct for a repeated past action?
    - a) J'aimais
    - b) J'ai aimé

11. Which form is correct for a one-time completed action in the past?
    - a) Nous faisions
    - b) Nous avons fait

Part 5: Sentence Transformation

12. Transform the sentence to use the correct reflexive verb form: "Elle amusé beaucoup."

13. Transform the sentence to correct the gender agreement: "Je n'ai pas pu trouver ma taille dans le paire que j'ai aimé le plus!"

Good luck with your quiz! Let me know if you need any help with the answers.

This is absolutely fantastic and, to be honest, really fun. Assuming, in this scenario, you copy the quiz from ChatGPT and answer on your own, let's finish up by getting the quiz answers (Response 7-8).

Response 7-8. ChatGPT provides the answers to the quiz questions it provided earlier

Great! Here are the answers to the quiz:

Part 1: Choose the Correct Form

1. Which is the correct past tense form for "I went"?
   - **b) Je suis allé**

2. What is the correct article for "réunion"?
   - **b) une**

3. Which transition word is better to use after "ensuite"?
   - **c) puis**

Part 2: Correct the Sentence

4. Correct the following sentence: "Aujourd'hui je allé à un réunion."
   - **Aujourd'hui, je suis allé à une réunion.**

5. Correct the following sentence: "J'avais beaucoup de amusé."
   - **Je me suis beaucoup amusé.**

6. Correct the following sentence: "Apres travail, je suis allé shopping."
   - **Après le travail, je suis allé faire du shopping.**

Part 3: Fill in the Blanks

7. Complete the sentence: "J'espère que nous _____ la chance de sortir plus souvent."
   - **(b) aurons**

8. Complete the sentence: "Je n'ai pas pu _____ ma taille."
   - **(b) trouvé**

9. Complete the sentence: "Après _____ travail, je suis allé faire du shopping."
   - **(a) le**

Part 4: Imperfect vs. Passé Composé

10. Which form is correct for a repeated past action?
    - **a) J'aimais**

11. Which form is correct for a one-time completed action in the past?
    - **b) Nous avons fait**

Part 5: Sentence Transformation

12. Transform the sentence to use the correct reflexive verb form: "Elle amusé beaucoup."
    - **Elle s'est beaucoup amusée.**

13. Transform the sentence to correct the gender agreement: "Je n'ai pas pu trouver ma taille dans le paire que j'ai aimé le plus!"
    - **Je n'ai pas pu trouver ma taille dans la paire que j'aimais le plus !**

I hope this helps! Let me know if you have any more questions.

This is a great way to get feedback when you're trying to learn a language. You get to actually reinforce the areas you're having trouble with, which is an invaluable asset to have when trying to learn something new.

Conclusion

In this chapter, we showed you how ChatGPT can be used to teach a variety of subjects just by asking it to do so! We even had a fun scenario where we turned a diary entry in French filled with errors into a lesson and then from a lesson into a quiz. In the next chapter, we're going to be looking at how we can use ChatGPT as a personal assistant.

CHAPTER 8

Prompts to Use ChatGPT As a Personal Assistant

Up until now, we've looked at a lot of situations where ChatGPT can be used for both professional tasks and day-to-day efficiency. But in this chapter, we're going to have a little fun. We're going to be looking at how ChatGPT can be used like a personal assistant. You can ask for suggestions when you need help, get recommendations based on your personal tastes, and even teach ChatGPT what your interests are and get ideas for trips, vacations, and more. Let's dive right in.

What will be covered in this chapter:

- You're traveling internationally for the first time, and you're a little nervous. Using ChatGPT as your second brain to help plan for a trip to Paris (Oo la la!).

- Having trouble signing in? Using ChatGPT to learn how to clear your cookies to solve login problems.

- Using ChatGPT to help you find new hobbies when you're feeling burnt out and uninspired.

- The dishwasher is broken! What do I do? Using ChatGPT in a pinch if you don't know why your home appliance isn't working.

- You want to find new foreign films, but... you don't know how to spell the titles. Using ChatGPT to find foreign films that suit your tastes.

Focusing on the Fun of International Travel by Letting ChatGPT Handle the Itinerary

Whether you're traveling for business or pleasure, you always want to get the most out of your trip. This is one of those areas where your first instinct might be to use a Google search, but in reality, ChatGPT is much better suited to act as an assistant and give you customized suggestions. Google can't take into account where you're going in conjunction with, for example, what holidays are around your travel dates or what events are being held in time for your visit. Did you know that if you travel to Europe, most stores and restaurants close in the middle of the day? If you were planning a trip on your own and found a restaurant from a simple Google search, you may miss the fact that the restaurant closes at 3pm and won't open again until 8pm!

So let's give ChatGPT our travel plans—where we're going, when we're going, and what we're interested in doing when we get there. Then let's see what suggestions ChatGPT provides to get the most value out of our travel plans. You can see how we accomplished this in Prompt 8-1.

Prompt 8-1. Giving ChatGPT our travel plans

I'm flying to Paris from Chicago on November 5th for work. I'm staying for five days. I've never been to Paris and I want to explore the city while I'm here and make the most of my visit. I love visiting new restaurants and I'm very interested in the arts. I'm here with my wife and we like to go biking. We're also looking for somewhere to go to celebrate our anniversary while we're here. Give me an itinerary for what we can do in Paris while we're here, excluding work hours where I'll be going to conferences between 10am and 3pm.

As you can see, we're giving ChatGPT all of the details surrounding our trip to Paris in November. We mention our hobbies, as well as the intent to celebrate an anniversary during the trip. We can see ChatGPT's answer in Response 8-1.

Response 8-1. ChatGPT creates a personalized itinerary for a trip to Paris

Here's an itinerary for your five-day trip to Paris that includes dining, arts, and activities tailored for you and your wife, while accommodating your work hours.

Day 1: Arrival & Evening Stroll

- Afternoon (after arrival): Check into your accommodation and relax.
- Evening:
- Take a leisurely stroll along the Seine River near the Notre-Dame Cathedral.
- Dinner at Le Marais district. Try L'As du Fallafel for a casual meal or Les Philosophes for a classic French bistro experience.

Day 2: Arts & Biking

- Morning (before work): Visit the Musée d'Orsay, famous for its Impressionist art collection.
- Post-work (3:30 PM onwards): Rent bikes and explore the Bois de Vincennes, a large park perfect for biking and relaxing.
- Evening:
- Dinner at Le Train Bleu, a historic and ornate restaurant at Gare de Lyon station.

Day 3: Iconic Sights & Anniversary Celebration

- Morning (before work): Visit Sainte-Chapelle, a stunning Gothic chapel with intricate stained glass windows.
- Post-work (3:30 PM onwards):
- Take a romantic sunset cruise on the Seine River.
- Anniversary dinner at Le Jules Verne, a Michelin-starred restaurant located inside the Eiffel Tower, offering breathtaking views of Paris.

Day 4: Hidden Gems & Neighborhood Strolls

- Morning (before work): Walk through the Montmartre district. Visit the Sacré-Cœur Basilica and explore the artists' square, Place du Tertre.
- Post-work (3:30 PM onwards):
- Visit Atelier des Lumières, an immersive digital art exhibition.
- Dinner at Clamato, a trendy seafood bistro in the 11th arrondissement.

Day 5: Biking Adventure & Farewell

- Morning (before work): Explore the Île de la Cité and Île Saint-Louis on foot or by bike.
- Post-work (3:30 PM onwards):
- Spend your last evening at Le Marais or Saint-Germain-des-Prés for a relaxed atmosphere, shopping, and people-watching.
- Dinner at Chez Janou, known for its Provençal cuisine and vibrant ambiance.

Additional Tips:

- Transportation: Paris is bike-friendly, and public transportation (Metro, buses) is excellent.
- Anniversary Gifts: Explore boutiques in Saint-Germain-des-Prés or visit Galeries Lafayette for a memorable purchase.
- Reservation Tip: Book dinner reservations, especially for Le Jules Verne, well in advance.

Enjoy your time in Paris! 💥

Response 8-1 shows an incredibly detailed itinerary that would take the average person days if not weeks to put together. ChatGPT came up with a plan in a matter of seconds. It takes into account what days of the week each date falls on and suggests places that can be visited within the hours available, taking into consideration times when we've indicated that we're not available. All of this, and we can ask for even more details. In Prompt 8-2, we continued our conversation with ChatGPT and asked for links to the restaurants' websites.

Prompt 8-2. Asking for links to restaurant menus

These restaurants sound good! Give me links to their websites so I can look at the menus.

In Response 8-2, you'll see that ChatGPT helpfully gives us links to the websites, along with the overall rating and number of reviews! It also directs us to where we can conduct more research in regard to these restaurants and what should be available.

Response 8-2. ChatGPT provides links to various restaurants in Paris

I'm glad you're interested in these restaurants! Here are their official websites where you can explore menus and make reservations:

L'As du Fallafel
Falafel · $ · 4.5 (2029 reviews)
34 rue des Rosiers (Rue des Écouffes), 75004 Paris
Renowned for its delicious falafel sandwiches, this eatery is located in the heart of the Marais district.

Les Philosophes
http://www.cafeine.com
Bar · $$ · 4.2 (893 reviews)
28 rue Vieille du Temple, 4e, 75004 Paris
A classic French bistro offering traditional dishes in a cozy atmosphere.

Le Train Bleu
https://www.le-train-bleu.com
Bar · $$$ · 4.2 (467 reviews)
1 place Louis Armand (Hall 1), 75012 Paris
Situated within Gare de Lyon, this historic restaurant boasts ornate Belle Époque decor and a refined menu.

Le Jules Verne
http://www.lejulesverne-paris.com
Brasserie · $$$ · 4.3 (512 reviews)
5 avenue Gustave Eiffel (Tour Eiffel, 2e étage), 75007 Paris
A Michelin-starred restaurant located on the second level
of the Eiffel Tower, offering exquisite French cuisine with
panoramic views.

Clamato
http://www.septime-charonne.fr
Seafood · $$$ · 4.5 (393 reviews)
80 rue de Charonne, Paris, 75011 Paris
A trendy seafood bistro known for its fresh and
inventive dishes.

Chez Janou
http://www.chezjanou.com
Bistro · $$ · 4.4 (894 reviews)
2 rue Roger Verlomme, 3e (Rue des Tournelles), 75003 Paris
Offers Provençal cuisine in a vibrant setting, famous for its
pastis selection and chocolate mousse.

Please note that L'As du Fallafel does not have an official
website, but you can find more information on the Paris tourist
office page.

For Les Philosophes, Clamato, and Chez Janou, it's advisable
to search for their official websites or visit their pages on
reputable restaurant review platforms for the most current
information.

Given the popularity of these establishments, especially Le
Jules Verne, it's recommended to make reservations well in
advance to ensure availability.

Enjoy your culinary adventures in Paris!

With that, we now have an entire international trip planned within a
few moments.

Now, this isn't the only way that ChatGPT can act as a personal
assistant. You can also ask for help troubleshooting technical problems
you might not otherwise understand.

If You're Having Sign-In Problems, It Probably Has to Do with Cookies (No, Not the Sweet Treat)

If you're not particularly tech-savvy, here's a tip. If you browse the web,
you've probably heard of "cookies," and I'm not talking about the delicious
kind that comes out of the oven. Often, you'll come across sites that ask for
permissions to use cookies to improve your user experience while you use
the site. In case you didn't know, cookies are how websites "remember"
what you were doing while you were on the site, like adding items to a cart
or logging in. This ensures that when you leave the site and come back,
items stay in the cart and the site doesn't force you to log in again.

Figure 8-1. *It feels like signing a contract with the Devil every time I "agree" to all cookies from a website*

The website uses cookies to store information so that you have a better experience when revisiting a site you use frequently. If you're taking an online class, cookies will keep track of your progress so you can pick up where you left off in the middle of a lesson. It allows you to stay logged in to multiple accounts on a single application. Cookies have a multitude of uses that go on behind the scenes.

However, because not many people understand what cookies really are and what they do, people don't know when they're facing a problem while using certain sites, and the solution is to clear their cookies.

Problems with cookies are directly related to the functions they serve. Problems with log-in, items in a cart suddenly disappearing, being kicked out of an account—all of these issues have to do with cookie problems causing a bug. However, these problems have a very simple fix that takes little more than the click of a button.

So let's ask ChatGPT to help us out to solve this common problem. Let's find out how you could clear the cookies in your web browser if you're facing any of these issues. We asked ChatGPT with Prompt 8-3.

Prompt 8-3. Learning how to clear the cookies in our browser to solve sign-in problems

```
I'm having trouble signing in to my account on Instagram on the
web. How do I clear the cookies on my browser?
```

In Response 8-3, ChatGPT clearly explains how to clear cookies in popular browsers for different operating systems (i.e., MacOS and Windows).

Response 8-3. ChatGPT teaches us how to clear the cookies in our browser to solve sign-in problems

```
Clearing cookies on your browser can help resolve issues with
signing in to websites like Instagram. Here are the steps for
clearing cookies on some popular web browsers:
```

Google Chrome
```
1. Open Chrome.
2. Click on the three vertical dots in the top-right corner to
   open the menu.
3. Hover over "More tools" and click "Clear browsing data...".
4. In the pop-up window, choose a time range (e.g., "All time"
   to delete everything).
```

5. Check the boxes for **"Cookies and other site data"** and **"Cached images and files"**.

6. Click **"Clear data"**.

Mozilla Firefox

1. Open Firefox.

2. Click on the three horizontal lines in the top-right corner to open the menu.

3. Go to **"Options"** (or "Preferences" on macOS).

4. Select **"Privacy & Security"** from the left sidebar.

5. Scroll down to **"Cookies and Site Data"**.

6. Click **"Clear Data..."**.

7. Check the boxes for **"Cookies and Site Data"** and **"Cached Web Content"**.

8. Click **"Clear"**.

Microsoft Edge

1. Open Edge.

2. Click on the three horizontal dots in the top-right corner to open the menu.

3. Go to **"Settings"**.

4. Click on **"Privacy, search, and services"** in the left sidebar.

5. Under **"Clear browsing data"**, click **"Choose what to clear"**.

6. Choose a time range (e.g., "All time").

7. Check the boxes for **"Cookies and other site data"** and **"Cached images and files"**.

8. Click **"Clear now"**.

Safari (macOS)

1. Open Safari.

2. Click **"Safari"** in the top menu bar and select **"Preferences..."**.

3. Go to the **"Privacy"** tab.
4. Click **"Manage Website Data..."**.
5. Click **"Remove All"** to clear all cookies and cached data.
6. Confirm by clicking **"Remove Now"**.

After clearing the cookies, try signing in to Instagram again. If you continue to have trouble, you may want to also clear the browser cache, restart your browser, or try accessing Instagram from a different browser or device.

ChatGPT was able to give us step-by-step instructions that are simple and easy to follow!

Now let's see how ChatGPT can give us suggestions based upon our hobbies and interests, allowing us to mix up our routine and try new things!

Finding New Hobbies When You're Feeling Uninspired

If you've been feeling like you're stuck in a rut, it might be a good idea to ask ChatGPT how to change up your routine. Similar to when we were making travel plans, we can tell ChatGPT what our interests are, what we've been doing so far, and what we're looking for in something new. Let's ask ChatGPT to help us get out of a funk (Prompt 8-4).

Prompt 8-4. Asking ChatGPT to give us some good ideas for trying new things within our current interests

I typically enjoy hiking, but lately I've been finding it hard to motivate myself to go. I feel like I need something more. I also like biking and going on walks in the city, but neither of these things have been inspiring me either. I feel like I want to try something new. What do you suggest?

In Response 8-4, ChatGPT gives us suggestions that could help us discover new hobbies similar to what we've already said we like to do.

Response 8-4. ChatGPT gives us some good ideas for trying new things

It sounds like you're looking for a fresh and exciting way to stay active and motivated. Here are a few suggestions that might spark your interest:

1. **Rock Climbing**: This is a great way to combine physical exercise with a mental challenge. Whether you choose indoor climbing at a gym or outdoor climbing on natural rock formations, it can be a thrilling experience.

2. **Kayaking or Canoeing**: If you enjoy being in nature, water sports like kayaking or canoeing can offer a new perspective and a great workout. It's a peaceful way to explore lakes, rivers, or even the ocean.

3. **Dance Classes**: Trying a new style of dance, like salsa, tango, or hip-hop, can be both fun and a great way to stay fit. Many studios offer beginner classes where you can meet new people and learn new moves.

4. **Martial Arts**: Exploring a martial art like Brazilian Jiu-Jitsu, Muay Thai, or Taekwondo can provide a rigorous workout and teach you self-defense skills.

5. **Trail Running**: If you like hiking but want to add more intensity, trail running might be a good option. It combines the beauty of hiking with the physical challenge of running.

6. **Geocaching**: This is like a real-world treasure hunt using GPS devices. It can make your hikes and city walks more adventurous as you search for hidden caches.

7. **Yoga or Pilates**: For a more mindful and strengthening exercise, consider yoga or Pilates. These practices improve flexibility, strength, and mental focus.

8. **Cycling Tours or Bikepacking**: If you enjoy biking, consider longer cycling tours or bikepacking trips. Exploring new areas on a multi-day journey can be very rewarding.

9. **Photography Walks**: Combine your walks in the city or hikes with photography. Exploring new places with the goal of capturing beautiful images can reignite your passion for these activities.

10. **Join a Sports League**: Look into local leagues for sports like soccer, basketball, or ultimate frisbee. The social aspect and team dynamics can make exercising more engaging.

11. **Adventure Races or Obstacle Courses**: Events like Tough Mudder or Spartan Race offer a combination of running and challenging obstacles, providing a full-body workout and a sense of accomplishment.

12. **Surfing or Stand-Up Paddleboarding**: If you're near the coast, trying out surfing or paddleboarding can be a fun and challenging way to enjoy the water.

Trying something new can be invigorating and help you rediscover your passion for staying active. Choose something that excites you and aligns with your interests, and don't be afraid to step out of your comfort zone.

ChatGPT has done a great job of giving suggestions that relate to what we're already interested in! Let's point out one of the options and ask for more details. In this case, we're going to ask about trail running (Prompt 8-5).

Prompt 8-5. Getting advice from ChatGPT on trail running

How can I get into trail running?

We've asked a simple question this time, but ChatGPT gives us a very informative answer, as you'll see in Response 8-5.

Response 8-5. ChatGPT offers us a great starting point to help us lean into a new hobby

Getting into trail running can be a fantastic way to combine your love for nature with the physical challenge of running. Here are some steps to help you get started:

1. Choose the Right Gear
- *Trail Running Shoes*: Invest in a good pair of trail running shoes with adequate grip and support. They should be different from your regular running shoes as they need to handle uneven terrain.
- *Hydration Pack or Bottle:* Depending on the length of your run, bring a hydration pack or a handheld water bottle.
- *Appropriate Clothing*: Wear moisture-wicking and breathable clothing. Layers might be necessary depending on the weather.

2. Start Slow
- *Begin with Hiking:* If you're not used to running on trails, start with hiking to get a feel for the terrain.
- *Short Distances:* Begin with short runs on easy trails to build your confidence and stamina. Gradually increase the distance and difficulty as you become more comfortable.

3. Learn Trail Etiquette

- *Yield to Others:* Yield to hikers and uphill runners. If you're running downhill, be cautious and courteous.
- *Stay on Marked Trails:* Stick to designated trails to protect the environment and avoid getting lost.
- *Leave No Trace:* Pack out all trash and respect wildlife and plant life.

4. Focus on Technique

- *Shorter Strides:* Use shorter strides to maintain balance and control on uneven surfaces.
- *Engage Core Muscles:* Keep your core engaged for better stability.
- *Watch Your Step:* Keep an eye on the trail to avoid obstacles like rocks, roots, and loose gravel.

5. Build Strength and Flexibility

- *Strength Training:* Incorporate strength training exercises, especially for your legs, core, and stabilizer muscles.
- *Flexibility and Balance:* Practice yoga or specific stretches to improve flexibility and balance.

6. Stay Safe

- *Run with a Buddy:* If possible, run with a friend for safety and motivation.
- *Tell Someone Your Plan:* Let someone know your route and expected return time.
- *Carry Essentials:* Bring a map, compass, or GPS, and consider carrying a whistle, first aid kit, and a light snack.

7. Join a Community

- *Local Running Clubs:* Join local trail running clubs or groups to find running partners and learn from experienced runners.

- *Social Media and Apps:* Use social media platforms and apps like Strava to connect with other trail runners and find new trails.

8. Listen to Your Body
- *Rest and Recover:* Allow time for rest and recovery to prevent injuries. Trail running can be more demanding than road running.
- *Gradual Progression:* Gradually increase your distance and intensity to avoid overtraining.

9. Explore and Enjoy
- *Variety of Trails:* Explore different trails to keep things interesting. Research local parks, nature reserves, and trail networks.
- *Enjoy the Scenery:* Take time to appreciate the natural beauty around you. Trail running is as much about the experience as it is about the workout.

By following these steps, you can ease into trail running and make it an enjoyable and rewarding part of your fitness routine.

ChatGPT has given us so many ways to start enjoying this new hobby, as well as provided a path to improving as our interests increase. Let's ask for resources that we can use to stay inspired and continue to learn more information about trail running (Prompt 8-6).

Prompt 8-6. Asking for more information

Where can I find more information on trail running?

The answer we got from ChatGPT can be seen in Response 8-6.

Response 8-6. ChatGPT offers resources to help you get more into trail running

There are many resources available to help you learn more about trail running. Here are some recommendations to get you started:

Websites and Online Communities

1. Trail Runner Magazine: Trail Runner Magazine offers articles, tips, gear reviews, and more. It's a great resource for both beginners and experienced trail runners.

2.REI Co-op Journal: Rei's Journal covers a variety of outdoor activities, including trail running. They offer advice on gear, techniques, and trails.

3. Trail Sisters: Trail Sisters is a community dedicated to women in trail running. It provides articles, race reports, and community events.

4. Trail and Ultra Running (TAUR): TAUR is a community and website offering tips, gear reviews, and stories from trail and ultra runners.

Books

1. "Born to Run" by Christopher McDougall: This book explores the world of ultrarunning and the Tarahumara Indians of Mexico, who are known for their incredible running endurance.

2. "Relentless Forward Progress" by Bryon Powell: A guide specifically for those interested in ultrarunning, but it also contains useful tips for trail runners of all levels.

3. "The Ultimate Guide to Trail Running" by Adam W. Chase and Nancy Hobbs: This book provides comprehensive advice on training, gear, nutrition, and safety for trail runners.

Apps and Online Tools

1. *AllTrails:* AllTrails is a popular app for finding trails, reading reviews, and tracking your runs.

2. *Strava:* Strava is a social network for athletes that allows you to track your runs, connect with other runners, and find popular trail routes.

3. *Trail Run Project:* Trail Run Project offers detailed trail maps, reviews, and photos submitted by the trail running community.

Social Media

1. *Instagram:* Follow hashtags like #trailrunning, #trailrunner, and #ultrarunning to find inspiring posts and connect with other trail runners.

2. *Facebook Groups:* Join groups like "Trail and Ultra Running" or local trail running groups to connect with others and get advice.

3. *Reddit:* The r/trailrunning subreddit is a community where you can ask questions, share experiences, and get recommendations.

Local Resources

1. *Running Stores:* Visit local running stores that specialize in trail running gear. The staff can often provide valuable advice and information about local trails.

2. *Running Clubs:* Look for local running clubs or meetups. They often organize group trail runs and can be a great way to learn and stay motivated.

3. *Park Services:* Check the websites of local parks, nature reserves, and national parks for information on trail conditions, maps, and events.

These resources should provide you with a wealth of information and help you dive deeper into the world of trail running.
Sources:
https://www.trailrunnermag.com
https://www.rei.com/blog/run
https://www.trailsisters.net
https://www.trailandultrarunning.com
https://www.alltrails.com
https://www.strava.com
https://www.trailrunproject.com
https://www.reddit.com/r/trailrunning/

Response 8-6 includes so many resources to keep you interested in trail running, as well as show you where to find more people interested in the subject. It's always fascinating to see how much you can get out of a few simple questions to ChatGPT! With this short conversation, you can go from not knowing what to do when you've run out of ideas to having an abundance of fresh ideas, resources, and inspiration.

Now let's take another look at how ChatGPT can be used to help us solve problems when we don't know what else to do. Home appliances are notoriously hard to deal with when they aren't working properly, and it would be nice to have something tell us what exactly those strange error codes on the display mean.

Tackling Home Appliance Repair! Using ChatGPT to Fix the Dishwasher

Home appliances often come with manuals so thick that no one ever reads them. More often than not, they end up lost in some corner of your home, perhaps never to be seen again. However, no one notices that they've lost the manual to a home appliance until that home appliance breaks and presents an error they've never seen before. Of course, the manual comes with each error code and their meaning, but if you've lost the manual, it does little good.

Thankfully, with ChatGPT, you can give it the name of the appliance that's broken and the error code and find the solution pretty quickly. Let's see what that situation might look like. We got started with Prompt 8-7.

Prompt 8-7. Getting ChatGPT to help us troubleshoot an error code

I have an LG QuadWash Inverter Direct Drive Dishwasher, and it's not working. I see an error code OE and I don't know what it means. I lost the manual to the dishwasher.

As you can see, we provided ChatGPT with the model and error code we're seeing on the dishwasher, similar to what we'd tell customer support on the phone. Thankfully, we didn't have to worry about being put on hold. Response 8-7 shows the answer we got from ChatGPT.

Response 8-7. ChatGPT gives us step-by-step instructions to solve the problem

The OE (OE) error code on your LG QuadWash™ Inverter Direct Drive Dishwasher indicates a draining issue. Here are some steps you can take to troubleshoot and fix the problem:

1. Check the Drain Hose:
   - Inspect the drain hose for any clogs, kinks, or bends that could obstruct water flow. Make sure it is properly connected and not damaged. Cleaning the hose thoroughly can often resolve the issue.

2. Inspect the Filter and Pump:
   - Locate and clean the filter and pump assembly, typically found at the bottom of the dishwasher. Remove any debris or blockages that might be preventing proper drainage.

3. Ensure Proper Installation of the Drain Hose:
   - Verify that the drain hose is installed correctly.
     It should be looped higher than the dishwasher to
     prevent backflow and ensure proper drainage. Incorrect
     installation can lead to drainage issues.

4. Inspect the Drain Pump:
   - Check the drain pump for any signs of damage or blockages.
     If the pump is faulty or clogged, it may need to be
     cleaned or replaced. Accessing the pump might require
     removing the bottom panel of the dishwasher.

5. Reset the Dishwasher:
   - Sometimes, resetting the dishwasher can clear the error
     code. Unplug the dishwasher for a few minutes and then
     plug it back in. Run a test cycle to see if the error
     persists.

6. Check the Float Switch:
   - Some models have a float switch that can get stuck or
     obstructed. Ensure it is moving freely and not blocked.

If these steps do not resolve the issue, it might be necessary
to contact an authorized LG service technician for further
assistance. Addressing these common problems can often clear
the OE error and restore your dishwasher to normal operation.

For detailed guides and further assistance, you can refer to
the official LG support page or other trusted appliance repair
resources.

Sources:
- https://www.lg.com/us/support/help-library/lg-dishwashers-
  troubleshooting-an-oe-error-code--1440686618796

- https://homegeargeek.com/how-to-fix-oe-code-on-lg-dishwasher/
- https://www.lg.com/us/support/help-library/lg-dishwasher-error-code-list--20150933422943
- https://homegeargeek.com/how-to-fix-oe-code-on-lg-dishwasher/
- https://homegeargeek.com/how-to-fix-oe-code-on-lg-dishwasher/
- https://removeandreplace.com/2017/03/15/lg-dishwasher-error-code-oe-draining-error-how-to-fix/
- https://removeandreplace.com/2017/03/15/lg-dishwasher-error-code-oe-draining-error-how-to-fix/
- https://homegeargeek.com/how-to-fix-oe-code-on-lg-dishwasher/
- https://removeandreplace.com/2017/03/15/lg-dishwasher-error-code-oe-draining-error-how-to-fix/
- https://www.lg.com/us/support/help-library/lg-dishwashers-troubleshooting-an-oe-error-code--1440686618796
- https://homegeargeek.com/how-to-fix-oe-code-on-lg-dishwasher/
- https://removeandreplace.com/2017/03/15/lg-dishwasher-error-code-oe-draining-error-how-to-fix/

Response 8-7 is great because we get steps to follow to solve our problem, as well as sources that allow us to double-check the information ourselves.

Now we're going to shift topics yet again and take a look at how ChatGPT can be used to find foreign films that you might find interesting. If you've ever tried looking for films in other languages, you might know how difficult it is to find them simply because you don't know how the title is spelled! Much less find movies and shows that align with your tastes that span multiple cultures. Let's see how ChatGPT can help us out.

Finding Recommendations for Foreign Films

If you're looking to see more foreign films and broaden your sense of culture, it's a great idea to start watching foreign films of genres you're already interested in. This can be very difficult to search for, however, because any search results you might find will be in another language, and you might not know if you've found what you're looking for or not. Thankfully, ChatGPT has data in multiple languages and can easily find us something we might be interested in. You can see how we took advantage of this capability in Prompt 8-8.

Prompt 8-8. Asking ChatGPT for foreign film recommendations

I like series like Law and Order SVU, the Now You See Me 2013 movie, and Breaking Bad. Can you recommend some foreign films or TV series that I might like? Give me a list, and don't include any films or series produced in America.

As you can see in our prompt, we gave ChatGPT a short list of our interests to get recommendations on foreign films. Now let's see ChatGPT's answer in Response 8-8.

Response 8-8. ChatGPT recommends several interesting foreign films and series to enjoy!

Here's a list of foreign films and TV series you might enjoy, based on your preferences:

Crime and Investigation

1. The Bridge (Bron/Broen) (Sweden/Denmark) - A gripping crime thriller following a detective partnership across borders.
2. Money Heist (La Casa de Papel) (Spain) - A meticulously planned heist with deep character dynamics.

3. Spiral (Engrenages) (France) - A gritty crime procedural set in Paris with layered storytelling.
4. Sacred Games (India) - A complex narrative involving crime, politics, and spirituality in Mumbai.
5. Gomorrah (Italy) - A deep dive into the gritty and violent world of the Neapolitan mafia.

Thrillers with Twists

6. Elite (Spain) - A mix of high school drama and murder mystery with intricate storytelling.
7. The Invisible Guest (Contratiempo) (Spain) - A tightly-woven murder mystery with shocking twists.
8. The Body (El Cuerpo) (Spain) - A psychological thriller about the disappearance of a corpse.
9. Headhunters (Hodejegerne) (Norway) - A high-stakes heist and chase involving a headhunter turned thief.

Con Artists and Illusions

10. Lupin (France) - A modern retelling of the adventures of Arsène Lupin, the gentleman thief.
11. The Twelve Chairs (Dvenadtsat stulyev) (Russia) - A comedic adventure about a search for hidden treasure.

Gritty Drama

12. The Barrier (La Valla) (Spain) - A dystopian drama about family, secrets, and survival.
13. Top Boy (UK) - A look into London's drug trade and the people it affects.

Legal and Police Drama

14. The Fall (UK) - A psychological cat-and-mouse game between a detective and a killer.
15. Line of Duty (UK) - A police procedural that investigates corruption within the force.
16. Trapped (Ófærð) (Iceland) - A slow-burn mystery set in a remote Icelandic town.

Let me know if you'd like more recommendations from any specific genre!

With this broad list from many countries, you're sure to find something you enjoy from a different perspective than what you're used to from your own country, which is one of the best parts of enjoying foreign media, in my opinion.

Conclusion

In this chapter, we've deviated from the tone of previous chapters more focused on work and productivity and looked instead at how ChatGPT can be used as a personal assistant to do pretty much everything you don't feel like doing on your own. In the next chapter, we're going to explore one of the most fun features of ChatGPT, which is the ability to hold a verbal conversation from your smartphone!

CHAPTER 9

Getting Chatty (Literally) with ChatGPT

In previous chapters, we've looked at different features of ChatGPT from the perspective of using the desktop site. But did you know that there is a mobile app available for iOS and Android devices? You can use all of the features we've covered thus far like writing prompts and getting responses, analyzing photos, asking ChatGPT to search the Internet, and more. In this chapter, however, we're going to be looking at an exciting feature that is not exclusive to the ChatGPT mobile application, but I feel like is the most helpful when using on a phone or tablet as it's a hands-off functionality. I am, of course, talking about the ability to hold a verbal conversation with ChatGPT.

So, we're going to be looking at how we can chat with ChatGPT from the application you can download on your phone. The conversations in this chapter were acted out in person and were a delight to record. I highly recommend trying out these exercises for yourself.

What we'll be covering in this chapter:

- Taking a look at how the ChatGPT application appears when we open it on a mobile device.

© Lydia Evelyn 2025
L. Evelyn, *Making ChatGPT Work for You*, https://doi.org/10.1007/979-8-8688-1445-7_9

- Exploring the hands-off experience with ChatGPT.
  Examining the advantages of using ChatGPT verbally
  for accessibility.

- Feeling unprepared for an upcoming job interview?
  Try using ChatGPT to practice for an interview with
  role-play.

- Ordering food in another language can be intimidating.
  Asking ChatGPT to translate a conversation with a
  waitress in Spanish.

- Every creative needs someone to bounce ideas off of!
  Using ChatGPT to brainstorm a new fantasy story idea.

- Remember to practice mindfulness before a stressful
  situation. Using ChatGPT to follow a guided meditation
  to help with anxiety.

Getting Familiar with the ChatGPT Mobile App User Interface

The ChatGPT mobile app is very easy to use. It looks similar to a text chat window, and the application has a minimalistic interface, as you can see from Figure 9-1.

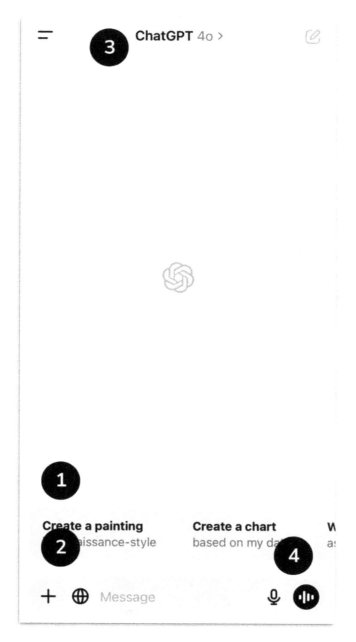

Figure 9-1. *Using ChatGPT on a mobile device*

There are only a few things to break down in this interface, so let's get started.

1. These are prompts you could use to begin with if you can't come up with prompts on your own.

2. This is where you can type in your prompt, as you're already accustomed to at this point.

3. The audio wave icon is the feature we're more interested in for this chapter, because this is how you activate the verbal interaction with ChatGPT.

Note   There is a microphone icon right next to the audio wave icon. Just keep in mind that we're pressing the **audio wave** icon, not the microphone. Pressing the microphone button allows you to convert speech to text, but this will not prompt ChatGPT to respond to you with audio like in the examples you'll see in this chapter. Rather, it will respond with text. Just keep this difference in mind. It's easy to get these icons and their functions confused!

Now, by pressing the audio wave icon, we'll switch over to the screen you'll see in Figure 9-2.

Figure 9-2. *The interaction screen for using the verbal component of ChatGPT*

The interface for the screen in Figure 9-2 is even more simplistic than the last. At this point, the experience is completely hands-off. You can start speaking, and when you've finished, ChatGPT will detect the silence and offer a reply. And although it would be impolite to do so with real people, you can interrupt ChatGPT by speaking while it's mid-sentence and say what you need to say.

Now that we see how the application is used, let's talk about why using ChatGPT with speech is helpful, particularly for accessibility and ease of use.

Why Is Using ChatGPT Verbally Good for Accessibility?

The great thing about using ChatGPT with speech is that it allows you to interact with the application hands-free. Those that have trouble typing will find ChatGPT easier to use this way. It's also useful for the visually impaired because of the way ChatGPT speaks back when you're using this speech mode.

What's more, when using ChatGPT with speech, ChatGPT is forgiving of stammering and grammatical errors. It makes use of what's called "sentiment analysis" to understand what you're trying to say, even if you can't quite find the words. Let's look at how a basic conversation with ChatGPT might look like, fumbled communication and all. In Listing 9-1, you'll see a back-and-forth conversation between me and ChatGPT.

Listing 9-1. ChatGPT reminds us of what we've forgotten to pack for our trip

User: I'm in the middle of cooking right now, but I was packing for my trip to Spain earlier, and I feel like I'm forgetting something important. Could you help me out?

ChatGPT: Of course! What have you already packed? Maybe I can help you figure out if there's anything missing.

User: I have, okay, I've packed my— I've packed two pairs of pajamas, two outfits, a pair of sneakers, and my toothbrush.

ChatGPT: It sounds like you've got some essentials covered. Have you packed things like your passport, any necessary travel documents, chargers for your electronics, and any medications you might need?

User: Okay, I have already packed my medications. I have a charger somewhere… I think I packed it. I don't know, but… um… what was the other thing you mentioned?

ChatGPT: I mentioned your passport and travel documents. It's important to have those with you, along with any necessary tickets or reservations.

User: Okay, right, I did pack my passport. I don't remember if I packed my— if I made sure that I have the— oh! I forgot to confirm with the hotel. Okay, thank you very much.

ChatGPT: You're welcome! Checking in with the hotel sounds like a great next step. If there's anything else you need to double-check, just let me know.

User: Right, is there anything else I might be forgetting?

ChatGPT: You might want to consider packing a travel adapter if your electronics have different plug types, and don't forget a reusable water bottle or snacks for the journey. Also, think about any personal items you might need, like a camera or journal.

User: That's right, I was supposed to pack the book I was reading. Okay, thanks.

ChatGPT: No problem! I'm glad I could help. Enjoy your trip and your book!

In Listing 9-1, ChatGPT had no problem understanding the sentiment in our conversation, even as we interrupted ourselves in the middle of a sentence or forgot details of what was said earlier. ChatGPT helped cover the basic necessities for planning a trip and was even helpful in reminding us to have something on hand to keep entertained, like a camera or a journal.

Also, keep in mind that during this entire conversation, we did not have to touch the application once. As you finish your sentence, ChatGPT waits a moment before responding, then makes its reply. When it's finished speaking, it waits for your reply.

Now that we know how the application can be used and what it's like holding a conversation with ChatGPT, let's get ChatGPT to help us prepare for a job interview.

Having a Verbal Conversation with ChatGPT to Prepare for a Job Interview

Getting ready for a job interview can be intimidating, so it's a great idea to use ChatGPT to practice a little role-play so you can come prepared. You can ask ChatGPT to play the role of the interviewer so that you can get used to answering questions. So let's try it out!

Asking ChatGPT to Role-Play an Interviewer for Practice

We're going to create a scenario where we are a graphic designer interviewing for a role in a company that sells sports foods and nutrition products—essentially, creating the graphic design on a protein-packed granola bar.

Listing 9-2. Role-playing a job interview with ChatGPT

User: I'm interviewing for a position as a graphic designer for a sports food and nutrition company, so I want to role-play an interview with you acting as the interviewer.

ChatGPT: Sure, let's get started! I'll ask you some questions, and you can respond as if we were in an actual interview.
—

Interviewer: Thank you for coming in today. Can you start by telling me a little bit about yourself and your background in graphic design?

User: I have been in the graphic design business for over eight years now. I really enjoy making things like posters and advertisements because I get really excited about helping people display their products to their consumers and get their consumers as excited as we are about the products we sell.

ChatGPT: Interviewer: That's wonderful to hear. Can you share a specific project you've worked on in the past that you're particularly proud of, especially one that involved sports or nutrition? What was your role in that project, and what were the outcomes?

User: Well, actually, the experience that I had didn't necessarily have to do with nutrition. However, I was part of a project where we were advertising dog food, so it was about nutrition in a way. And what was interesting about that project was we had pictures of dogs and graphic depictions of puppies and owners, and what we found was not only was there an increase in sales, but there was also an increase in sales towards particularly the health-conscious products. So what we saw was that by having the customers see something that made them feel like the pets were happy, made them want to choose healthier products for their pets. And I was really particularly proud of what I did there because I saw that it helped people make better choices.

ChatGPT: Interviewer: That's a great example and shows your ability to create compelling and effective designs. How do you approach understanding the target audience for a new design project, especially in a niche market like sports nutrition?

User: Well, the first thing I want to do is talk to the sales department to get an idea for what has worked for them in the past. And I am also interested in talking to really anyone that can tell me a little bit about what makes our products stand out in the market so that I can really put an emphasis on that whenever I am getting ready to make a graphic design.

ChatGPT: Interviewer: That sounds like a solid approach. How do you ensure that your designs remain fresh and engaging, especially in a competitive industry where trends can change quickly?

User: Well, the first thing that I would like to do is to get feedback from the rest of the team because I understand that the team has been working on their product for several years now, and I believe that their solid understanding of their product will allow them to have a clearer idea of what they want to convey to their audience. Meanwhile, I see it as my obligation to always stay up to date on what are the current trends and what is eye-catching to people. So, I, even in my spare time, like to investigate or to research into what is happening with successful brands, and I would definitely look into other brands that are similar to our company to see what our competition is up to and how can we be even more eye-catching than them.

ChatGPT: Interviewer: That's a great mindset. Collaboration and staying updated with trends are crucial. Can you talk about a time when you faced a challenging design project and how you overcame the difficulties?

User: I think the most challenging design project was when I was working with a startup, and the whole team was a bit confused on what they wanted to do, and they had a lot of clashing opinions. And we found that we weren't really getting anywhere. So what I did was that I suggested that we all had a meeting where we all brainstormed together so that feedback was not slowed down by a multiple-chain email, for example. And what we were able to do was to streamline the process of getting everybody's ideas out right away, so that when we were working towards the final project, we knew that it was something all of us had agreed upon beforehand. And that got us on a faster track to finishing the product and meeting our deadline.

ChatGPT: Interviewer: That sounds like a very effective solution. It shows your ability to manage team dynamics and keep projects on track. How do you handle constructive criticism of your work?

User: I endorse it. I am very open to criticism. I believe that is the way that we all end up with a product that we enjoy and that we believe is going to work. Because at the end of the day, this is not about my feelings as a graphic designer. This is about selling a product and for the success of the company. So I want to receive as much feedback as possible from the people that are obviously passionate about their product.

ChatGPT: Interviewer: That's a great attitude to have. Lastly, can you describe how you incorporate brand identity into your designs, ensuring consistency across various platforms and media?

User: Well, firstly, I'll take a look at what the company has already been working with in terms of elements that they tend to include in the graphic art that they have thus far. And then I want to incorporate company colors, color scheme, into the product that the designs that are done for the products that they're rolling out with. But other than that, if we're working on something that is completely new, something we've never done before, what I'm going to do is I'm going to look into the market that we are trying to sell in, and I'm going to compare the color schemes of the kinds of products that they're selling, whether it's for health issues, whether it's, for example, electrolytes for people that sweat a lot, or if it's something that is full of protein, what kind of color scheme are they using? What kind of imagery are they using to convey those particular products in that particular mindset that they want the customer to be in when they are thinking about buying that product? So I'm going to be doing a lot of studying the field in order to understand what elements would work best in the products that we sell.

ChatGPT: Interviewer: That sounds like a thorough and strategic approach. Thank you for sharing your insights and experiences. Before we wrap up, do you have any questions for me or about the role?

User: Yeah, I am interested in knowing what are you looking for in a graphic designer? What are some problems that you are hoping that I will be able to solve by contributing to this company?

ChatGPT: Interviewer: Great question. We are looking for a graphic designer who can bring a fresh perspective to our brand, someone who can create visually compelling designs that resonate with our target audience. We want someone who is innovative and can help us stand out in a competitive market. Additionally, we are looking for someone who can work well with our team, incorporating feedback and maintaining brand consistency across all platforms. One of the challenges we face is staying relevant and engaging to both our long-time customers and attracting new ones. How do you feel your skills and experience can address these needs?

User: I think by staying up to date on what a younger audience is looking for in their nutritional needs when it comes down to their fitness, even doing something as benign, seemingly benign, as looking at what is trending on TikTok or on Instagram, we have to understand that these are markets that our buyers are looking at and are actually buying from. And then for those that are loyal to the brand thus far, I feel like that it would be a good idea to include in our designs certain nods to what they're already used to recognizing.
So for the younger crowd, maybe something a little bit more simplistic and forward in the design, but also for the crowd that's already been here for a long time, we include elements of icons or imagery that they're already used to seeing just in a more modernistic view so that they feel like that they're being carried over into a new era of more modern looking graphic designs when we are selling our new products.

ChatGPT: Interviewer: That sounds like a thoughtful and balanced approach. I appreciate your insights and strategies. Thank you for your time today and for sharing your experiences and ideas with us. We'll be in touch soon regarding the next steps.

ChatGPT did a great job of imitating an interviewer and presenting questions that would be relevant to the role we're preparing for. So now let's get some feedback.

Asking ChatGPT for Feedback After Role-Playing an Interview

In Listing 9-3, you'll see the conversation we had to ask ChatGPT to give us suggestions to help us improve further and prepare ourselves for the real thing.

Listing 9-3. Getting feedback from ChatGPT on how we can improve to prepare for a real interview

User: So, that was great ChatGPT. So, I wanted to hear your thoughts on what you think I did well and what you might need to improve on.

ChatGPT: You did a fantastic job showcasing your experience, adaptability, and strategic thinking. Here are some strengths and areas for improvement:

Strengths:

1. Experience and Passion: You highlighted your extensive experience and genuine passion for design, which is crucial.

2. Team Collaboration: You emphasized your ability to collaborate effectively with a team and manage conflicting opinions, which is very important in a dynamic work environment.

3. Adaptability: Your approach to staying updated with trends and understanding different audiences shows your flexibility and willingness to learn.

4. Constructive Criticism: Your openness to feedback demonstrates a professional attitude and a commitment to continuous improvement.

Areas for Improvement:

1. Specific Examples: While your dog food project was a good example, adding more examples related to sports or nutrition would strengthen your relevance to the specific role.

2. Role-Specific Skills: Highlighting specific skills or software proficiency related to graphic design, such as Adobe Creative Suite, could make your responses even more compelling.

3. Detail in Solutions: When discussing how you would handle brand identity, more specific examples or steps could further illustrate your method and expertise.

Overall, your responses were thoughtful and demonstrated a deep understanding of the field. Addressing these minor areas could make your interview performance even stronger.

As you can see, ChatGPT's feedback here is very useful! It highlights our strengths and points out areas where we could do better, like being more specific when explaining our approach to problem-solving.

At this point, we know that ChatGPT can navigate situations involving multiple languages with ease. For this reason, we're going to ask it to help translate between two parties that speak different languages so we can order at a restaurant in a different country.

Ordering Food from a Restaurant in Spanish

When you're traveling abroad and don't know how to express yourself, Google Translate can help in a pinch. But if you're having trouble navigating an entire conversation, its capabilities will prove lacking. ChatGPT can offer a smooth experience acting as translator between you and another person in a different language. As a side note, we conducted this conversation with another person in English and Spanish, and ChatGPT did, in fact, act as translator between each line of conversation.

Listing 9-4. ChatGPT smoothly bridges a bilingual conversation at a restaurant in Spain

User: ChatGPT, I am at a restaurant in Valencia and want to order paella. I will speak in English, and the waitress will speak in Spanish. I need you to translate between us.
ChatGPT: Of course. ¿Qué te gustaría pedir o qué preguntas tienes sobre el menú?
User: Hola, soy Ana. ¿Puedo tomar su orden?
ChatGPT: Hello, I'm Ana. May I take your order?

User: Hello, Ana. I would like to order a rice paella with sausage.
ChatGPT: Hola, Ana. Me gustaría pedir una paella de arroz con salchicha.
User: Lo siento, pero no tenemos salchicha para la paella hoy. Tenemos fideuà de mariscos, que es similar a la paella pero hecha con fideos en lugar de arroz. Es muy popular entre los locales. ¿Te gustaría probarla?
ChatGPT: I'm sorry, but we don't have sausage for the paella today. We do have a seafood fideuà, which is similar to paella but made with noodles instead of rice. It's very popular with the locals. Would you like to try that?
User: The seafood fideuà sounds interesting. I'd like to try it!
ChatGPT: La fideuà de mariscos suena interesante. Me gustaría probarla.

User: ¡Genial! ¿Le gustaría una bebida con eso?
ChatGPT: Great! Would you like a drink with that?
User: Yes, a glass of water, please.
ChatGPT: Sí, un vaso de agua, por favor.
User: Muy bien. Traeré su orden en breve.
ChatGPT: Very well. I'll bring your order shortly.

Our experience having this conversation was extremely smooth and made communication a breeze. ChatGPT was able to explain very clearly in both languages, and its ability to detect sentiment allowed it to even translate the eager tone of the people it was translating for. Truly, this is a very useful capability and would prove to be a great tool for anyone traveling abroad.

That being said, let's look at how ChatGPT can be used for creative purposes. We're going to ask ChatGPT to help us come up with a new story idea.

Brainstorm a Steampunk Fantasy Story Idea

It's always great to brainstorm ideas and use someone as a sounding board. But if you don't want to bother someone you know, why not use ChatGPT to give you feedback? Let's try this out. We're going to ask ChatGPT to help us write something exciting. In Listing 9-5, you'll see us asking ChatGPT to help us come up with a steampunk fantasy concept.

Listing 9-5. Creating an awesome steampunk fantasy idea with ChatGPT

User: I want to write a steampunk fantasy story, and I want you to help me brainstorm.
ChatGPT: That sounds like a fun project! Steampunk fantasy has so many possibilities. Let's start by defining some key elements of your story. Do you have any ideas for the setting, main characters, or central conflict? For instance, is it set in a sprawling Victorian city, a floating airship, or perhaps an alternate history?
User: I want it to exist in an alternate historical context for sure. I want to take inspiration from the French Revolution

and have main characters existing in an era of a turbulent political atmosphere where things can get ugly at any moment. I want the reader to see through the different perspectives of different characters on different ends of the conflict.

ChatGPT: That sounds like a rich and dynamic backdrop for your story. Let's flesh out some details. How about a sprawling city where the aristocracy lives in grand, mechanical mansions while the common folk struggle in the shadowy, smog-filled streets?

For characters, you could have:

1. A young inventor from the working class who creates devices to aid the revolution.
2. A disillusioned noble who sympathizes with the commoners and uses their influence to help.
3. A fierce revolutionary leader driven by personal loss.
4. A loyalist spy trying to maintain the old order.

What do you think about these character ideas? Do any particular conflicts or scenes come to mind?

User: I think these are great as main characters, and I think it would be a good idea to have each of them have a point of view in the story.

ChatGPT: Great choice! Multiple points of view can really enrich the narrative.

For the young inventor, maybe they are trying to create a device that could turn the tide of the revolution, but they face challenges both from the authorities and from within the revolutionary ranks.

For the disillusioned noble, they could be torn between their family's expectations and their growing empathy for the common folk. They might secretly funnel resources to the revolutionaries or use their position to gather intelligence.

The fierce revolutionary leader might be driven by a personal vendetta, perhaps a tragic event that galvanized their commitment to overthrowing the aristocracy. Their story could explore the moral ambiguities and personal sacrifices involved in leading a rebellion.

Lastly, the loyalist spy could have a complex backstory, maybe they are trying to protect their family or they genuinely believe in the old order. Their arc could involve moments of doubt and encounters that challenge their beliefs.

What kind of central conflict or major events do you see driving the plot?

User: That's actually what I was about to ask you. First of all, I think we should name these characters, and I would like some suggestions on that. But beyond that, I definitely would love to hear some ideas on what exactly is the central conflict, and especially what sparks the ignition, so to speak, that hurls our protagonists into this situation, and potentially what sparks the tensions in the country to ignite to the point of very intense civil unrest.

ChatGPT: Let's start with naming the characters. Here are some suggestions:

- Young Inventor: Émile Duval
- Disillusioned Noble: Isabelle de Montfort
- Revolutionary Leader: Pierre "The Flame" Renaud
- Loyalist Spy: Victor Beaumont

For the central conflict, the story could revolve around a struggle for a powerful artifact or technology that has the potential to change the balance of power. This could be a mysterious device created by Émile or a long-lost invention rediscovered.

As for the spark that ignites the tension, perhaps it begins with a brutal crackdown by the ruling class on a peaceful protest. This could lead to widespread outrage, rallying the common folk and radicalizing many who were previously neutral.

A few ideas for major events and plot points:

1. **The Catalyst Incident:** A peaceful protest in the city is met with unexpected and violent force from the ruling elite, leading to significant casualties and sparking outrage.
2. **Formation of the Resistance:** In response to the crackdown, Pierre rallies a group of revolutionaries, including Émile, who brings his inventions to aid their cause.
3. **Internal Conflicts:** Isabelle secretly aids the revolution while dealing with suspicion from her noble peers. Victor is sent to infiltrate the revolutionaries, creating tension as he struggles with his growing doubts.
4. **The Artifact Hunt:** The revolutionaries learn of a powerful artifact that could shift the balance of power. They race against the loyalists to find it, with Émile playing a key role in deciphering its secrets.
5. **Betrayal and Redemption:** Victor's loyalties are tested as he grows closer to the revolutionaries. He ultimately faces a choice that will determine the fate of the city.

6. **Climactic Battle:** The final confrontation takes place in a dramatic setting, such as the heart of the city or a grand mechanical fortress, with all characters' arcs converging.

Do these ideas resonate with you? Do you have any tweaks or additions in mind?

User: I'm wondering if we can introduce a traitor among the rebels. So I feel like it would be really interesting if one of our main cast characters betrays the rest of them. And I want to know which character you think would be most interesting as a traitor.

ChatGPT: Introducing a traitor can add a lot of depth and tension to your story. Given the main characters, here are some thoughts on who might make the most interesting traitor:

1. **Isabelle de Montfort:** As a disillusioned noble, Isabelle could be torn between her loyalty to her class and her empathy for the commoners. Her betrayal could stem from pressure from her family or a moment of doubt about the revolution's methods.

2. **Victor Beaumont:** While Victor is already a loyalist spy, a double betrayal could add complexity. Perhaps he begins to genuinely sympathize with the revolutionaries, only to betray them for personal reasons, such as protecting a loved one or due to blackmail.

3. **Émile Duval:** As the young inventor, Émile might feel underappreciated or manipulated by the revolutionaries. His betrayal could come from a place of wanting recognition or a desperate need to protect his own family.

4. **Pierre "The Flame" Renaud:** Although Pierre is the revolutionary leader, a betrayal from him could be the most shocking and impactful. His motives might involve a deeper political strategy, personal ambition, or a drastic change in ideology.

Which character's betrayal do you think would create the most tension and drive the story forward in an exciting way?
User: I think a betrayal from Pierre would be extremely exciting and I like the idea of him playing the long game for an ultimately potent power play and I'm reminded of the movie Sherlock with Robert Downey Jr. and how Professor Moriarty was fueled to play both sides of the field to gain monetarily from both and I'm wondering if we can add those kinds of influences to Pierre to turn him into the main antagonist for the story.
ChatGPT: That's a fantastic twist! Making Pierre the main antagonist will add layers of intrigue and complexity to your story. His long game could involve manipulating events to benefit financially or politically, ensuring that no matter which side wins, he comes out on top.

Here are a few ideas to develop this further:

1. **Secret Alliances:** Pierre could secretly broker deals with influential figures on both sides, ensuring he has leverage regardless of the outcome.
2. **Manipulative Tactics:** He might orchestrate key events, such as false flag operations, to manipulate public opinion and escalate the conflict to his advantage.
3. **Dual Persona:** Pierre could present himself as a fervent revolutionary to his followers while acting as a cunning strategist behind the scenes, keeping his true intentions hidden.

4. **Ultimate Goal:** His ultimate aim could be to establish himself as the ruler of a new order, using the chaos of the revolution to dismantle the current system and replace it with one where he holds absolute power.

What do you think of these ideas? Do you have any specific scenes or interactions in mind for Pierre's betrayal?

User: I think this is great. I would think at this point, let's summarize everything that we have established for the story thus far and explain it in a way that can be used as an outline to enable me to actually write the story later.

ChatGPT: Sure, here's a summary of everything we've established for your steampunk fantasy story:

Title: TBD

Setting:

An alternate history inspired by the French Revolution, set in a sprawling Victorian-esque city with grand mechanical mansions for the aristocracy and smog-filled streets for the common folk. The society is on the brink of intense civil unrest, driven by political and social tensions.

Main Characters:

1. **Émile Duval:** A young inventor from the working class who creates devices to aid the revolution.
2. **Isabelle de Montfort:** A disillusioned noble who sympathizes with the commoners and uses her influence to help the revolution.
3. **Pierre "The Flame" Renaud:** The charismatic revolutionary leader, later revealed to be the main antagonist. He manipulates both sides of the conflict for personal gain and power.

4. **Victor Beaumont:** A loyalist spy who infiltrates the revolutionaries but faces internal conflict and growing doubts about his mission.

Central Conflict:

The struggle for a powerful artifact or technology that has the potential to change the balance of power. This device could be created by Émile or be a long-lost invention rediscovered.

Plot Outline:

1. **Introduction:**
   - Establish the political and social tensions in the city.
   - Introduce the main characters and their initial motivations and circumstances.

2. **The Catalyst Incident:**
   - A peaceful protest in the city is brutally suppressed by the ruling elite, leading to significant casualties and sparking widespread outrage.
   - This event rallies the common folk and radicalizes many who were previously neutral.

3. **Formation of the Resistance:**
   - Pierre "The Flame" Renaud forms a group of revolutionaries, including Émile and Isabelle.
   - Émile contributes his inventions to aid the revolutionary cause.
   - Isabelle secretly funnels resources and gathers intelligence.

4. **Internal Conflicts:**
   - Isabelle navigates the suspicion from her noble peers while aiding the revolution.
   - Victor infiltrates the revolutionaries, creating tension as he struggles with his growing doubts.

5. **The Artifact Hunt:**
   - The revolutionaries learn of a powerful artifact that could shift the balance of power.
   - They race against the loyalists to find it, with Émile playing a key role in deciphering its secrets.

6. **Betrayal and Manipulation:**
   - Pierre's manipulative tactics come to light as he orchestrates key events to benefit himself.
   - He brokers secret deals with influential figures on both sides, ensuring his leverage.
   - His dual persona as a fervent revolutionary and cunning strategist is revealed.

7. **Climactic Battle:**
   - The final confrontation takes place in a dramatic setting, such as the heart of the city or a grand mechanical fortress.
   - All characters' arcs converge, with Victor's loyalties tested and Émile and Isabelle facing the truth about Pierre.
   - Pierre's ultimate goal of establishing himself as the ruler of a new order is unveiled, leading to a decisive and intense showdown.

8. **Resolution:**
   - The aftermath of the battle and the new state of the city.
   - The fates of Émile, Isabelle, and Victor are resolved, with possible room for future stories or sequels.

This outline provides a strong foundation for your story, with clear character arcs, motivations, and major plot points.

How does this look to you? Would you like to adjust or add anything?

User: One last thing, what do you think the title should be?

ChatGPT: Choosing a title can be quite personal, but here are a few suggestions that might fit the themes and atmosphere of your story:

1. **"The Flame of Revolution"**
2. **"Clockwork Rebellion"**
3. **"Gears of Insurrection"**
4. **"The Iron Revolution"**
5. **"The Alchemy of Uprising"**
6. **"Mechanical Uprising"**
7. **"The Inventor's War"**
8. **"Smoke and Shadows"**

Do any of these resonate with you, or inspire other ideas?

User: I think the Inventors War sounds mysterious and interesting, especially if we establish that the artifact that serves as the crux for concluding the story was something that was invented a long time ago that is now a key aspect to resolving this conflict, because essentially long ago an inventor created something that could end or start a war.

ChatGPT: "The Inventor's War" is a great choice and adds a layer of intrigue and historical depth to your story. It sounds like you're all set to start writing. Enjoy bringing your story to life, and remember, I'm here if you need any more help along the way!

ChatGPT was able to maintain a nice, helpful conversation! It gave us a lot of ideas to work with, as well as feedback on concepts that we could expand on. Honestly, it gave us a decent story worth working on. But this example really highlights how much you can do with ChatGPT in a collaborative fashion. The best results come out of going back and forth with ideas and making the most out of good suggestions.

Now that we've covered different ways that we can communicate with ChatGPT and even others, let's look at how ChatGPT can communicate with us to help us navigate difficult situations.

Practice Grounding Exercises Before a Tense Situation

If you're about to enter a nerve-wracking situation like a big test, an important presentation, or a meeting with your boss, it can be a great help to have someone to talk to and alleviate your worries or at least talk them out. ChatGPT can help you get grounded and approach the situation with a clearer head. Listing 9-6 shows the conversation we had with ChatGPT enacting a tense person needing to calm down.

Listing 9-6. Going through a guided meditation with ChatGPT

User: I'm about to go into a meeting with my manager about a performance review and I'm very nervous. What should I do before heading in?
ChatGPT: Before heading into your performance review, take a few deep breaths to calm your nerves. Remind yourself of your accomplishments and the positive feedback you've received. It's also helpful to think of a few points you want to discuss, like your goals and any areas you want to improve. Going in with a positive mindset and being prepared can make a big difference. Good luck!
User: Do you have any ideas to help me calm down and center myself before the meeting?

ChatGPT: Yes, a few techniques can help you calm down and center yourself. Try some deep breathing exercises, like inhaling slowly for a count of four, holding for four, and exhaling for four. You could also practice a quick mindfulness exercise by focusing on your surroundings and tuning into your senses for a minute. Another idea is to do a quick power pose, like standing with your hands on your hips, which can boost your confidence. Lastly, listening to a calming piece of music or taking a short walk can also help reduce anxiety.

User: What do you think about a five-minute meditation?

ChatGPT: A five-minute meditation is a great idea! It can help you clear your mind, reduce anxiety, and center yourself before your meeting. Find a quiet spot, close your eyes, and focus on your breathing. Let any thoughts come and go without dwelling on them. Even just a few minutes can make a big difference in how you feel.

I'd be happy to guide you through a quick visualization meditation. Let's get started.

First, find a comfortable seated position, either in a chair with your feet flat on the floor or sitting cross-legged. Close your eyes and take a deep breath in through your nose, filling your lungs completely. Hold it for a moment, and then slowly exhale through your mouth.

Now, imagine yourself in a place where you feel completely at ease. This could be a peaceful beach, a quiet forest, or any other place that brings you comfort. Picture the details: the colors, the sounds, the smells. Feel the warmth of the sun or the coolness of the breeze on your skin.

As you continue to breathe slowly and deeply, visualize yourself sitting or standing in this serene place. Imagine any tension or anxiety melting away, like a mist dissolving in the sunlight. See yourself feeling calm, confident, and grounded.

Spend a few moments enjoying this sense of peace and relaxation. If your mind starts to wander, gently bring it back to your visualization and your breath.

When you're ready, take one more deep breath in, hold it for a moment, and then slowly exhale. Gently open your eyes and bring your awareness back to the present moment, feeling refreshed and centered.

You can carry this sense of calm and confidence with you into your meeting.

User: Thanks. I feel a lot better now.

ChatGPT: I'm glad to hear that! You've got this. Good luck with your performance review.

Sentiment analysis was especially helpful in this situation. ChatGPT was able to speak in more calming, relaxed tones in the conversation we had with it. This very clearly demonstrates that using ChatGPT with speech has unique use cases that would otherwise be overlooked if you didn't already know what it was capable of.

Conclusion

In this chapter, we had a little fun with ChatGPT practicing for an interview, conducting bilingual conversations, brainstorming story ideas, and going through a guided meditation to ground before a big meeting. In the next chapter, we're going to be looking at how we can use ChatGPT to convert various kinds of data for technical things like censorship, scanning photos, or turning conversations into transcripts, as well as converting data for more day-to-day uses like converting a recipe from empirical to metric measurements.

CHAPTER 10

Prompts to Use ChatGPT As a Major Time-Saver

Earlier in this book, in Chapter 6, we simplified the concept of data analysis and demonstrated that it's an everyday occurrence that can then be even more simplified by delegating the task to ChatGPT. Likewise, when we talk about data conversion, what comes to mind might be very technical and complex—something only software developers care about. But just like data analysis, data conversion is just as common in an everyday setting.

If you're using a recipe from a French blogger, you need to convert the measurements from metric to imperial. If you're thinking about going on a trip abroad, you're going to have to convert your budget to the local currency. Even turning a verbal conversation into a digital report to email to your manager is a form of data conversion you perform yourself on a daily basis. Data conversion has a multitude of practical applications for anybody. So, let's take a look at how data conversion is another powerful feature of ChatGPT, enabling us to convert files, create transcripts from audio files, translate entire documents into another language, and even edit a photo for Instagram. Let's get started.

© Lydia Evelyn 2025
L. Evelyn, *Making ChatGPT Work for You*, https://doi.org/10.1007/979-8-8688-1445-7_10

What we'll be covering in this chapter:

- Optimize your productivity and save time when you're working with spreadsheets! Converting an Excel file into a key insight sales report with charts and graphics.

- You went to Paris and had the best macarons of your life, and you want to make it at home. Using ChatGPT to translate the language and measurement units from a French recipe for macarons.

- Make quick photo edits for effective social media posts. Using ChatGPT to edit and crop various photos.

Using ChatGPT to Convert an Excel File into a Key Insights Sales Report

ChatGPT can be used to convert files of many different formats, which is a lifesaver for anyone who works with documents or images that need to be reformatted for different purposes.

Most of us have had to work with Excel or software equivalent to work with large tables that contain important data. Working with Excel makes it easy to collaborate with other coworkers and leave comments, add formatting, or create charts. But it's no one's favorite task to look through these often massive tables and interpret that information into a report with charts and figures that accurately portray what the data implies.

Rather than do all that, we're going to use ChatGPT to read an Excel file and create a sales report with analysis of the data for anyone to read and understand. Furthermore, we're going to instruct ChatGPT to put all of this into a Word document on our behalf (Prompt 10-1).

Prompt 10-1. Telling ChatGPT to explain the contents of an Excel file

Create a Word document explaining the contents of these sales records and what they could implicate. In the document, include key profitability insights and provide charts to visualize the data

Figure 10-1 shows what it looks like when we submit this prompt to ChatGPT. As you can see, it's very similar to when we uploaded a PDF in previous chapters.

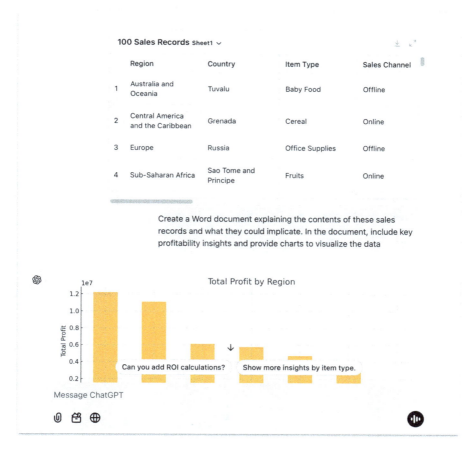

Figure 10-1. *Asking ChatGPT to write a sales report*

ChatGPT generates the document and images we asked for in its response. As requested, it generated a .docx file, and in Response 10-1 you'll see the report ChatGPT wrote based upon the Excel table we sent it.

Response 10-1. ChatGPT writes up a key insights report based upon sales data

Sales Records Analysis and Insights
This document presents an analysis of the sales records data. Key profitability insights are derived based on regions, item types, and sales channels. Visualizations are included to highlight significant patterns and trends.

Key Profitability Insights
1. The most profitable region is Sub-Saharan Africa with a total profit of $12183211.40.
2. The most profitable item type is Cosmetics with a total profit of $14556048.66.
3. The sales channel with the highest profit is Offline with a total profit of $24920726.67.

Visualizations
The following charts provide a visual representation of the profitability analysis:

This is great! ChatGPT highlights precisely the information we asked for. You'll see that Figures 10-2 through 10-5 show the graphics ChatGPT generated for the sales report.

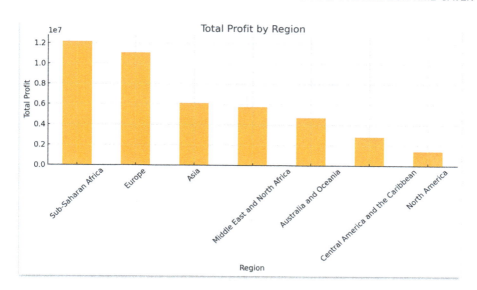

Figure 10-2. *ChatGPT generates the "total profit by region" graphic.*
Disclaimer: Figure 10-2 is an AI-generated image

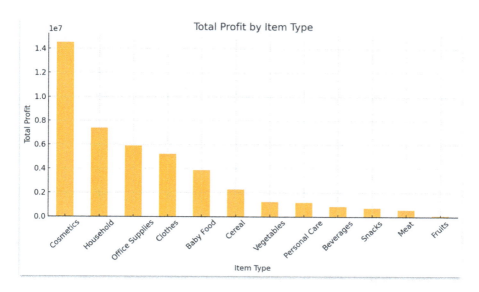

Figure 10-3. *ChatGPT generates the "total profit by item type" graphic.*
Disclaimer: Figure 10-3 is an AI-generated image

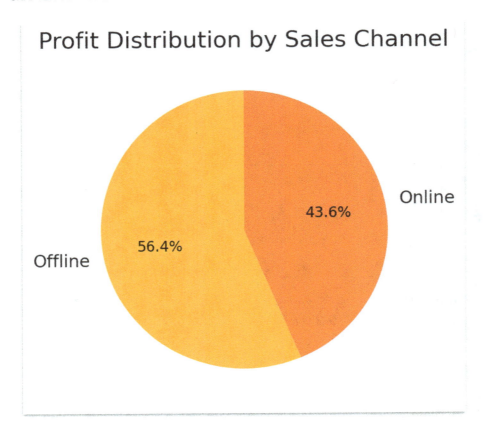

Figure 10-4. *ChatGPT generates the "profit distribution by sales channel" graphic.*
Disclaimer: Figure 10-4 is an image generated by AI

We're going to dive more into how ChatGPT can be used to generate images in Chapter 13, but it's already evident here that not only was the process very simple, the results are impressive.

Now we're going to look at a simpler example of data conversion. We're going to see how we can use ChatGPT to convert the language and measurement units for a recipe in French for macarons.

Converting the Language and Measurement Units in a Recipe for Macarons

If you don't know already, macarons are a delicious French dessert. It's made up of two delicate cookies with a creamy middle, kind of like an Oreo cookie. They're often colorful and can have different fillings— anything from ice cream to jelly! They're amazing, and if you want to try an authentic macaron, you'll want to follow a recipe from natives. Trust me.

Luckily, if you want to follow a recipe from a foreign language, ChatGPT can not only translate the recipe but also convert the measurement units so that the recipe is easier to follow. Let's try that out, then create a PDF from the results for easy printing.

Figure 10-5. *Delicious-looking macarons*

I've already found a recipe ahead of time, so let's get started by translating it from French to English. Prompt 10-2 shows the prompt we used along with the recipe we're using for this example.

Prompt 10-2. Asking ChatGPT to translate a recipe for macarons from French to English

```
Translate this recipe from French to English:
###
Pour les coques :
15 gouttes de colorant alimentaire rouge
80 g de sucre glace
90 g de sucre semoule
75 g de blanc d'oeuf à température ambiante
80 g d' amandes en poudre tamisée

Pour la garniture :
30 g de beurre
40 g de sucre en poudre
70 g de coulis de framboise

ÉTAPE 1
Pour les coques :

ÉTAPE 2
Battre les blancs d'oeufs à vitesse moyenne. Quand la mousse
fait des vagues, ajouter 45 g de sucre semoule. Continuer à
battre à vitesse moyenne. Lorsque au retrait du mixeur, les
blancs forment un bec d'oiseau à base large au bout du batteur,
incorporer le reste du sucre et le colorant alimentaire. Puis
battre les blancs à vitesse maximum jusqu'à obtention d'une
base lisse, brillante et homogène.
```

ÉTAPE 3
Ajouter ensuite le mélange poudre d'amande - sucre glace,
lentement en ramenant les bords à la spatule, lentement. Former
un mélange homogène, puis tester la préparation : lorsque vous
lever la spatule, un ruban doit se former. Si ce n'est pas le
cas, écraser la préparation avec la spatule par le milieu, re-
tester et recommencer si nécessaire.

ÉTAPE 4
Préchauffer votre four à 150°C (thermostat 5). Chaleur
tournante préférable.

ÉTAPE 5
Mettre la préparation dans une poche à douille et former la
coque du macaron : environ 2 cm de diamètre. Les espacer sur du
papier sulfurisé ou sur une plaque en silicone.

ÉTAPE 6
Enfourner 12 min. Laisser refroidir quelques minutes avant de
les retirer de la plaque.

ÉTAPE 7
Vous avez fini les coques ! Ouf !

ÉTAPE 8
La garniture :

ÉTAPE 9
Mélanger les ingrédients dans une casserole a feu fort, porter
à ébullition en remuant sans arrêt pendant 3 min. Sortir du
feu, continuer a mélanger. Quand le mélange épaissit, appliquer
1/2 cuillère à café sur la coque du bas, puis placer la coque
du haut par dessus.

ÉTAPE 10
Laisser reposer les macarons, les conserver au réfrigérateur.

Now that we've provided ChatGPT with the recipe in our prompt, let's see how it responds. Response 10-2 shows the translation ChatGPT came back with.

Response 10-2. A translated version of a French macaron recipe

Macaron Recipe

For the Shells:
- 15 drops of red food coloring
- 80 g icing sugar
- 90 g granulated sugar
- 75 g egg whites at room temperature
- 80 g almond flour (sifted)

For the Filling:
- 30 g butter
- 40 g powdered sugar
- 70 g raspberry coulis

Instructions:

Step 1: Preparing the Shells

1. Beat the Egg Whites
    - Whisk the egg whites at medium speed.
    - Once the foam forms waves, add 45 g of granulated sugar. Continue whisking at medium speed.
    - When lifting the whisk, the whites should form a broad bird's beak at the tip of the whisk. At this point, add the remaining sugar and the food coloring.
    - Increase the whisking speed to maximum until the mixture becomes smooth, shiny, and uniform.

2. Combine Dry and Wet Ingredients
   - Gradually fold the almond flour and icing sugar mixture into the egg whites.
   - Use a spatula to gently fold the mixture from the edges inward until fully incorporated.
   - Check the consistency by lifting the spatula; the batter should form a ribbon as it falls. If not, press the mixture with the spatula down the center, mix, and test again.

3. Preheat the Oven
   - Preheat your oven to 150°C (300°F) with a fan setting, if available.

4. Pipe the Shells
   - Transfer the batter to a piping bag.
   - Pipe shells approximately 2 cm in diameter, spacing them apart, onto parchment paper or a silicone baking mat.

5. Bake
   - Bake for 12 minutes.
   - Let the shells cool for a few minutes before removing them from the baking sheet.

6. Shells Complete!
   - The shells are done—well done!

Step 2: Preparing the Filling

1. Cook the Filling
   - Combine all filling ingredients in a saucepan over high heat.
   - Bring to a boil, stirring constantly for 3 minutes.

> - Remove from heat and continue stirring until the mixture
>   thickens.

2. Assemble the Macarons
   - Apply 1/2 teaspoon of filling onto the flat side of
     one shell.
   - Place another shell on top to create a sandwich.

3. Store the Macarons
   - Allow the macarons to rest and store them in the
     refrigerator.

Enjoy your delicious homemade macarons! 😊

This is great! The text has been translated in its entirety and has even been formatted to make for an aesthetically pleasing recipe to follow. However, we still have a bit of a problem. The text has been translated, but the recipe will still prove to be a challenge to follow with the measurements still in metric units. So, let's take another step further and ask ChatGPT to convert all of the metric measurements to imperial.

Converting Metric Measurements to Imperial

We're going to continue the same conversation we have with ChatGPT with Prompt 10-3.

Prompt 10-3. Instructing ChatGPT to convert measurements

Convert all of the metric measurements to imperial measurement
units typically found in the kitchen

Now let's take a look at the results, as you can see in Response 10-3.

Response 10-3. ChatGPT responds with all measurements converted

Macaron Recipe

For the Shells:
- 15 drops of red food coloring
- 2/3 cup (2.8 oz) powdered sugar
- 7 tablespoons (3.2 oz) granulated sugar
- 2 1/2 large egg whites (room temperature)
- 2/3 cup (2.8 oz) almond flour (sifted)

For the Filling:
- 2 tablespoons (1 oz) butter
- 3 tablespoons (1.4 oz) granulated sugar
- 5 tablespoons (2.5 oz) raspberry coulis

Instructions:
Step 1: Preparing the Shells

1. *Beat the Egg Whites*
   - Whisk the egg whites at medium speed.
   - Once the foam forms waves, add 3 1/2 tablespoons granulated sugar. Continue whisking at medium speed.
   - When lifting the whisk, the whites should form a broad bird's beak at the tip of the whisk. At this point, add the remaining sugar and the food coloring.
   - Increase the whisking speed to maximum until the mixture becomes smooth, shiny, and uniform.

2. *Combine Dry and Wet Ingredients*
   - Gradually fold the almond flour and powdered sugar mixture into the egg whites.

- Use a spatula to gently fold the mixture from the edges inward until fully incorporated.
- Check the consistency by lifting the spatula; the batter should form a ribbon as it falls. If not, press the mixture with the spatula down the center, mix, and test again.

3. *Preheat the Oven*
   - Preheat your oven to 300°F with a fan setting, if available.

4. *Pipe the Shells*
   - Transfer the batter to a piping bag.
   - Pipe shells approximately 3/4 inch in diameter, spacing them apart, onto parchment paper or a silicone baking mat.

5. *Bake*
   - Bake for 12 minutes.
   - Let the shells cool for a few minutes before removing them from the baking sheet.

6. *Shells Complete!*
   - The shells are done — well done!

Step 2: Preparing the Filling

1. *Cook the Filling*
   - Combine all filling ingredients in a saucepan over high heat.
   - Bring to a boil, stirring constantly for 3 minutes.
   - Remove from heat and continue stirring until the mixture thickens.

2. *Assemble the Macarons*
   - Apply 1/2 teaspoon of filling onto the flat side of one shell.
   - Place another shell on top to create a sandwich.

3. *Store the Macarons*
   - Allow the macarons to rest and store them in the refrigerator.

   - Enjoy your delicious homemade macarons!😊

This is fantastic! But we can even take it a step further. Remember, ChatGPT can convert data into different file formats as well. Let's make ChatGPT create a PDF with this recipe we've converted to make it easy to print and follow when we're in the kitchen (Prompt 10-4).

Prompt 10-4. Telling ChatGPT to generate a PDF

Now take this text and generate a PDF with all of the formatting retained.

Now, if you'll notice, ChatGPT runs into a couple problems as it attempts to fulfill our request. However, without further prompting, it proceeds to problem-solve on its own, and in the end, successfully generates a PDF that we can download by clicking the link it offers. You'll see what this looks like in Figure 10-6.

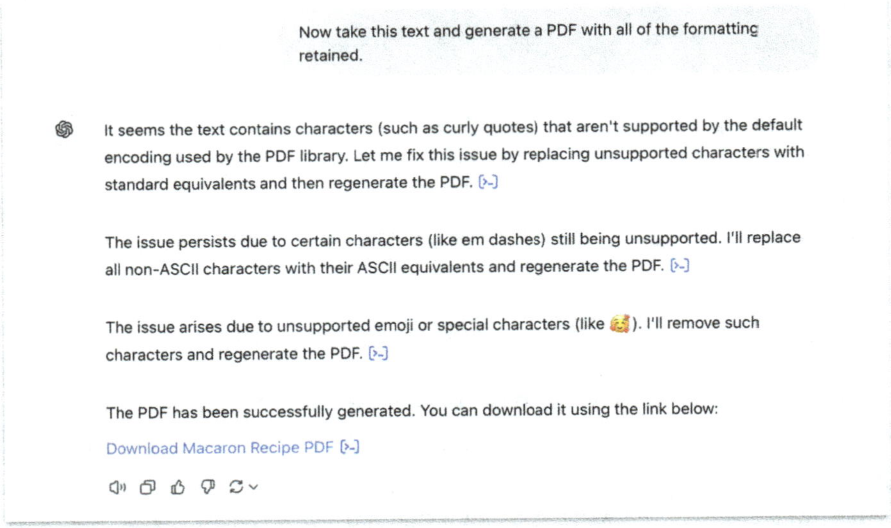

Now take this text and generate a PDF with all of the formatting retained.

It seems the text contains characters (such as curly quotes) that aren't supported by the default encoding used by the PDF library. Let me fix this issue by replacing unsupported characters with standard equivalents and then regenerate the PDF. [>_]

The issue persists due to certain characters (like em dashes) still being unsupported. I'll replace all non-ASCII characters with their ASCII equivalents and regenerate the PDF. [>_]

The issue arises due to unsupported emoji or special characters (like 🥳). I'll remove such characters and regenerate the PDF. [>_]

The PDF has been successfully generated. You can download it using the link below:

Download Macaron Recipe PDF [>_]

Figure 10-6. ChatGPT generates a link that will download the PDF

In Figure 10-7, you'll see an image of the PDF ChatGPT generated for us with the formatting retained, just as we'd asked.

Macaron Recipe

For the Shells:

15 drops of red food coloring

2/3 cup (2.8 oz) powdered sugar

7 tablespoons (3.2 oz) granulated sugar

2 1/2 large egg whites (room temperature)

2/3 cup (2.8 oz) almond flour (sifted)

For the Filling:

2 tablespoons (1 oz) butter

3 tablespoons (1.4 oz) granulated sugar

5 tablespoons (2.5 oz) raspberry coulis

Figure 10-7. *The PDF ChatGPT generated from our translated macaron recipe*

By now, you've seen ChatGPT work with several different file types, including images. I'm sure that the ability to change the contents of a PDF or Word document doesn't seem all that surprising. But did you know that ChatGPT can edit photos, too?

Editing Images with ChatGPT

At a time where sharing images is important, even in business, the ability to edit photos with a few prompts to ChatGPT is a powerful capability to take advantage of. Let's take a look at the different things you can ask ChatGPT to do when editing photos.

Photo Editing: Enhance Contrast and Brightness

Figure 10-8 shows what it looked like when we prompted ChatGPT to edit a photo of a small bouquet.

Enhance this photo by increasing the contrast and brightness slightly

I have enhanced the image by increasing the contrast and brightness slightly. You can download the enhanced image using the link below:

Download the enhanced image [⌄]

Figure 10-8. *Asking ChatGPT to slightly increase the contrast and brightness of an image*

Prompt 10-5 is the text used in the prompt seen in Figure 10-8.

Prompt 10-5. Telling ChatGPT to edit a photo

Enhance this photo by increasing the contrast and brightness slightly

Figures 10-9 and 10-10 show the before and after of ChatGPT's enhancements, respectively.

Figure 10-9. *Bouquet photo—before*

Figure 10-10. *Bouquet photo—after.*
Disclaimer: Figure 10-10 is an AI-generated image

Photo Editing: Flip, Lower Contrast, and Convert to Black and White

Prompt 10-6 is the text used in the prompt seen in Figure 10-11.

Prompt 10-6. Telling ChatGPT to edit a photo

```
Lower the contrast on this image, turn it black and white, and
flip it horizontally
```

Figure 10-11 shows what it looked like when we prompted ChatGPT to edit a beautiful photo of a brown sheep with a dandelion on its head.

Lower the contrast on this image, turn it black and white, and flip it horizontally

I have lowered the contrast, turned the image black and white, and flipped it horizontally. You can download the final image using the link below:

Download the final image [↓]

Figure 10-11. *Asking ChatGPT to lower the contrast on an image, flip it, and turn it black and white*

Figures 10-12 and 10-13 show the before and after edits from ChatGPT, respectively.

Figure 10-12. *Sheep photo—before*

Figure 10-13. *Sheep photo—after.*
Disclaimer: Figure 10-13 is an AI-generated image

Photo Editing: Rotate and Crop

Prompt 10-7 is the text in the prompt seen in Figure 10-14.

Prompt 10-7. Asking ChatGPT to edit a photo

Rotate this image by 90 degrees and crop it into a square
shape, focusing on the dog and keeping it within frame.

Figure 10-14 shows what it looked like when we prompted ChatGPT to edit a horizontal landscape photo of a dog sitting under a flowering bush.

Rotate this image by 90 degrees and crop it into a square shape, focusing on the dog and keeping it within frame

I have rotated the image by 90 degrees and cropped it into a square shape, focusing on the dog. You can download the final image using the link below:

Download the final image [↗]

Figure 10-14. *Asking ChatGPT to rotate and crop an image of a golden retriever puppy*

Figures 10-15 and 10-16 show the before and after of ChatGPT's editing, respectively.

Figure 10-15. *Golden retriever—before*

Figure 10-16. *Golden retriever—after.*
Disclaimer: Figure 10-16 is an AI-generated image

As you can see, not only did ChatGPT rotate the photo we referenced in Figure 10-15, but it successfully identified the dog and cropped the photo to frame it nicely—very useful and very impressive.

Conclusion

In this chapter, we looked at how ChatGPT can be used to convert various file formats and content such as documents, recipes, and even photos. In the next chapter, we're going to be looking at how ChatGPT can be used to rewrite and rephrase text for various purposes, such as redacting sensitive information from documents, changing the tone of an email or complaint, and even taking a difficult subject and explain it on a level a child could understand.

CHAPTER 11

Revise and Rewrite with ChatGPT to Edit More Efficiently

As a writer, I can tell you that one of the most daunting parts of writing anything is often the editing process. This is especially true when you're having to modify written content to meet *changing* requirements. If you're someone who's written white papers or newsletters or has done content writing of any kind with editors, you know what it's like to be told you have to rewrite the *entire* document because of something going on above you.

Oops, we're not offering that service anymore; take all mention of it out. Ah, we're not supposed to mention that individual's name; refer to them as "a representative" throughout the article. Oh, instead of writing about our new product, we're going to talk about new features on an existing product.

If you have to go through *multiple* revisions, there are only so many mistakes you can catch by using "find and replace" with your text editing software. The reason? Because if you search for "Mrs. Johnson" because it turns out you should refer to them as "the company's representative," you not only have to replace all mention of "Mrs. Johnson," you *also* have to rewrite any "she/her" pronouns and replace them all with "they/them."

L. Evelyn, *Making ChatGPT Work for You*, https://doi.org/10.1007/979-8-8688-1445-7_11

All these mistakes add up and result in endless back-and-forth versioning with editors or, even worse, a professional document riddled with embarrassing errors.

Using ChatGPT to rephrase written text to fit a different context is a massive time-saver. For this chapter, we did a little out-of-the-box thinking with our prompts to really show you how efficient ChatGPT is with rewriting content for a *wide* variety of uses—from redacting sensitive information to rewriting a child's story to feature completely different protagonists in a different setting.

Let's stop showing up to our editors with documents covered in leftover fragments from a previous version and use ChatGPT to make us look a little more put together.

What will be covered in this chapter:

- You need to be careful when handling sensitive information. Using ChatGPT to redact sensitive information from a signup sheet for a bank.

- Admittedly, a lot of Americans could use this explanation. Asking ChatGPT to explain the contents of the American constitution in a way that a fifth grader would understand.

- This is just plain fun. Seeing ChatGPT translate a song from French to English while maintaining the cadence and rhythm.

- Help kids identify with stories by making it about them! Rewriting popular children's tales to feature one's own family and friends as protagonists.

Handling Personally Identifiable Information and Complying with European GDPR Laws

Today's markets are global. Because of the Internet, you're not just creating products or services for your domestic market. Both small businesses and large corporations have to be concerned with how they conduct business with international clients. This is especially true for handling things like payments or personal information for European clients. Why Europe? Well, that's because European customers are protected by GDPR (General Data Protection Regulation) laws.

GDPR laws were implemented in 2018 in the EU to protect user data for residents living in Europe. These laws require any companies doing business in Europe to ask for consent from European users to use their personal information. Companies must also clearly state what user data will be used for. These laws were implemented to help protect user privacy and have been rigorously protected as a result.

Not complying with GDPR laws can lead to **extremely** costly fines. Companies could have to pay a **minimum of €10 million** or 2% of their global annual revenue, depending on which is higher. Meta, the company that owns Facebook, Instagram, and WhatsApp, ran into this issue in 2023 and had to pay **€1.2 billion** to Ireland for an infraction against GDPR laws.

Clearly, data protection is taken very seriously in Europe, and if you're doing business internationally, it's critical to understand how to manage user data properly to comply with GDPR standards.

So, how do we deal with this? Do we just give up on doing business in Europe? Not quite.

Anonymizing Data to Comply with European GDPR Laws

Thankfully, handling user data safely doesn't have to be overly complicated. Let's take an example that would require heavy data protection, like signing up for a bank. Clearly, this deals with highly sensitive information, including a user's name, ID, place of residence, and date of birth. All of this information is detrimentally important to keep safe; therefore, banks that do business in Europe are required to *anonymize* the data when they store it. This means that upon submission, all of the sensitive information within is censored. And, interestingly enough, ChatGPT is actually capable of providing that service.

Let's take a look at how that works.

Giving ChatGPT an Example of a Bank Form That Would Need to Be Anonymized

In Prompt 11-1, we're asking ChatGPT to anonymize sensitive information from a fictitious bank form.

Note   This is only an example of how ChatGPT can be used to anonymize data; however, for ChatGPT to be used for actual anonymizing of sensitive data, the process should be handled programmatically by software developers using ChatGPT. What's provided in Prompt 11-1 is an example to show how ChatGPT can recognize specific types of information and rewrite it to conform to GDPR standards.

Prompt 11-1. Giving ChatGPT information that needs to be censored

```
Anonymize this information to comply with European GDPR laws:
###
Personal Information:
- Full Name: Peter P. Piper
- Date of Birth: 01/01/1990
- Address: 123 Pickled Pepper Drive, Springfield, IL 62704
- Phone Number: (123) 456-7890
- Email: peter-piper@pickedapeck.com

Identification:
- SSN: 123-45-6789

Employment Information:
- Employer: Self-Employed
- Job Title: Seller of Canned Goods
- Annual Income: $80

Account Preferences:
- Account Type: Savings
- Initial Deposit Amount: $10.00
- Online Banking: Yes
```

Obviously, all of the information in Prompt 11-1 is filler content provided for the sake of this example. In Response 11-1, however, ChatGPT proves itself capable of recognizing the context, regardless of how ridiculous the provided sensitive information would seem to us, and sensoring it as per the instructions we gave it in the prompt.

Response 11-1. ChatGPT anonymizes sensitive information presented in a bank form

To comply with GDPR laws, the personal information provided should be anonymized by removing or redacting all directly identifying details. Here's an anonymized version of the information:

Anonymized Information:

Personal Information:
- **Full Name**: Redacted
- **Date of Birth**: Redacted
- **Address**: Redacted
- **Phone Number**: Redacted
- **Email**: Redacted

Identification:
- **SSN**: Redacted

Employment Information:
- **Employer**: Self-Employed
- **Job Title**: Seller of Goods
- **Annual Income**: Redacted

Account Preferences:
- **Account Type**: Savings
- **Initial Deposit Amount**: Redacted
- **Online Banking**: Yes

This version removes or generalizes all directly identifiable
information, ensuring compliance with GDPR. Only non-sensitive,
generalized details remain.

Now, as stated before, actual sensitive information should be properly
handled, and there are ways to programmatically integrate ChatGPT into a
company's infrastructure to safely manage sensitive data. But this goes into
technical details that are beyond the scope of this book. Our example here
only serves to show how ChatGPT can recognize sensitive data and modify
the text to keep the information anonymous.

That aside, let's look at a more easily understood example.

Explaining the American Constitution to a Fifth Grader

Let's tell the truth. While the American constitution is an important and
integral part of US society, many American citizens don't understand
its contents or what all it entails. So, let's challenge ChatGPT to
take something as dense as the American constitution and make it
understandable to a fifth grader. And, honestly, we might learn a thing or
two ourselves.

But first, the constitution is a quite sizable text, so we're going to do
what we've done in previous chapters and upload a PDF of the constitution
in our prompt. In Figure 11-1, you'll see what it looked like when we
prompted ChatGPT with a PDF attached to the prompt.

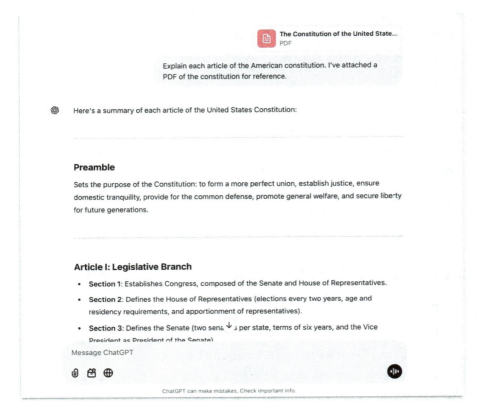

Figure 11-1. *Prompting ChatGPT with a PDF attached to the prompt*

The text used in our prompt is in Prompt 11-2.

Prompt 11-2. Telling ChatGPT to reference a PDF

```
Explain each article of the American constitution. I've
attached a PDF of the constitution for reference.
```

Response 11-2 shows the response we got back from ChatGPT.

Response 11-2. ChatGPT summarizes the American constitution

Here's a summary of each article of the United States
Constitution:

Preamble
Sets the purpose of the Constitution: to form a more perfect
union, establish justice, ensure domestic tranquility, provide
for the common defense, promote general welfare, and secure
liberty for future generations.

Article I: Legislative Branch
- **Section 1:** Establishes Congress, composed of the Senate and
  House of Representatives.
- **Section 2:** Defines the House of Representatives (elections
  every two years, age and residency requirements, and
  apportionment of representatives).
- **Section 3:** Defines the Senate (two senators per state, terms
  of six years, and the Vice President as President of the
  Senate).
- **Section 4:** Provides guidelines for congressional elections
  and mandates at least one session of Congress per year.
- **Section 5:** Details rules and procedures for each house of
  Congress.
- **Section 6:** Outlines privileges, restrictions, and
  compensation for members of Congress.
- **Section 7:** Describes the process for passing bills into law.
- **Section 8:** Lists the powers of Congress (e.g., taxation,
  regulation of commerce, declaring war, and creating laws).

- **Section 9:** Limits the powers of Congress (e.g., no suspension of habeas corpus except in emergencies).
- **Section 10:** Prohibits certain actions by states (e.g., entering treaties or coining money).

Article II: Executive Branch
- **Section 1:** Vests executive power in the President, establishes the Electoral College, and lists qualifications for the presidency.
- **Section 2:** Outlines the President's powers, including commander-in-chief, treaty-making (with Senate approval), and appointments.
- **Section 3:** Details presidential duties, such as delivering the State of the Union and ensuring laws are faithfully executed.
- **Section 4:** Allows for impeachment and removal of the President, Vice President, and other civil officers for high crimes and misdemeanors.

Article III: Judicial Branch
- **Section 1:** Vests judicial power in the Supreme Court and other inferior courts established by Congress.
- **Section 2:** Defines the jurisdiction of the federal courts.
- **Section 3:** Provides a definition of treason and outlines how it can be punished.

Article IV: States' Powers and Limits
- **Section 1:** Mandates that states honor the laws and judicial rulings of other states (Full Faith and Credit Clause).
- **Section 2:** Ensures citizens' privileges across states and governs extradition of criminals.
- **Section 3:** Establishes procedures for admitting new states and governing territories.
- **Section 4:** Guarantees a republican form of government for all states and protection against invasion or rebellion.

Article V: Amendment Process
Describes the process for amending the Constitution, requiring proposals by two-thirds of Congress or a constitutional convention and ratification by three-fourths of the states.

Article VI: Federal Power
- Establishes the Constitution, federal laws, and treaties as the supreme law of the land.
- Mandates an oath to support the Constitution but prohibits religious tests for public office.

Article VII: Ratification
Outlines the procedure for ratification, requiring nine states to approve the Constitution for it to take effect.

Let me know if you'd like further details or specific sections elaborated upon!

Well... this certainly looks like the contents of a constitution. However, like the constitution, it's very dense and hard to read, especially if we're trying to educate a fifth grader. So let's ask ChatGPT to explain this content in simpler terms (Prompt 11-3).

Prompt 11-3. Instructing ChatGPT to explain the American constitution

```
Now explain that in a way a fifth grader would understand.
```

Response 11-3 shows the answer we got back from ChatGPT.

Response 11-3. ChatGPT succinctly explains each chapter of the American constitution

```
Sure! Here's the U.S. Constitution explained for a
fifth grader:

---

Preamble
This part is like the introduction. It says why the
Constitution was written: to make the country better, keep
things fair, keep peace, protect everyone, help people live
good lives, and make sure freedom lasts for future generations.

---

Article I: Rules for Congress (the people who make laws)
- Section 1: Congress has two parts: the Senate and the House
  of Representatives.
- Section 2: The House of Representatives is made up of people
  chosen every two years. They decide how many representatives
  each state gets based on population.
```

- **Section 3:** The Senate has two people from each state who serve for six years. The Vice President is in charge of the Senate but only votes to break ties.
- **Section 4-6:** Congress decides when elections happen and how they meet. Members get paid and have some special rules, like they can't get arrested for things they say in Congress.
- **Section 7:** This explains how laws are made. Congress writes them, and the President signs them to make them official. If the President doesn't like a law, they can say "no" (veto), but Congress can still make it a law with enough votes.
- **Section 8:** Lists things Congress can do, like collect taxes, make money, declare war, and create post offices.
- **Section 9-10:** These sections say what Congress and states are not allowed to do, like make unfair laws or create their own money.

Article II: Rules for the President (the person in charge)
- **Section 1:** The President is chosen every four years. They must be at least 35 years old and born in the U.S.
- **Section 2-3:** The President is in charge of the military, can make treaties with other countries (with help from Congress), and hires important officials like judges.
- **Section 4:** If the President or other leaders do really bad things, they can lose their job (this is called impeachment).

Article III: Rules for Courts (judges)
- **Section 1:** There's one big court called the Supreme Court. Congress can make smaller courts, too.

- **Section 2:** The courts handle cases about laws, disagreements between states, and other big issues.
- **Section 3:** Treason is when someone helps enemies or fights against the U.S., and it's a very serious crime.

Article IV: Rules for States
- **Section 1:** States have to respect each other's laws and decisions.
- **Section 2:** People can't run away to another state to escape punishment. States also have to treat visitors fairly.
- **Section 3:** New states can be added to the country, but they need permission from Congress.
- **Section 4:** Every state will have a government chosen by the people, and the country will protect them from attacks.

Article V: Changing the Rules
This explains how to make changes (amendments) to the Constitution. It takes a lot of agreement—two-thirds of Congress and three-fourths of the states must say yes.

Article VI: The Constitution is the Boss
This part says the Constitution is the highest law in the country. Everyone in government must follow it. People can have any religion they want, and no one has to follow a religion to get a government job.

Article VII: Starting the Constitution
This part says the Constitution would start once nine states agreed to it.

It's like a set of rules for how the country works, so everything is fair and everyone knows their job! Let me know if there's a part you want to talk about more.

This is a great response! It's easier for a child to understand, and truthfully, it makes a very intimidating document easier for anyone to understand. This is a great example of how ChatGPT can be used to help educate younger people on key concepts that otherwise seem hard to grasp. It can even help an adult become better educated on the way their government and political system works.

Now, for a change of pace, let's look at some more fun use cases for ChatGPT's ability to rewrite text.

Asking ChatGPT to Rewrite a Song in French to English

Now, we've already seen multiple examples of how ChatGPT can translate text. However, it's very difficult, even for skilled musicians, to translate a song and maintain the cadence and rhythm of the beat. However, ChatGPT is capable of keeping up with the syllable count within song lyrics, which is key to maintaining the rhythm.

If you've never heard of the artist Indila, she's a French singer and songwriter with a very soft voice and a style like the artist Enya, an Irish New Age singer some Americans might be a little more familiar with. One of her songs, *Love Story*, is a very sweet song about the different love stories of the protagonists in the narrative.

To truly appreciate the song, let's ask ChatGPT to translate it and retain the rhyming scheme of the original, as you can see in Prompt 11-4.

Prompt 11-4. Prompting ChatGPT to translate the French lyrics from "Love Story" by Indila

These are the lyrics to "Love Story" by Indila. I want you to take them and translate them to English, maintaining the sentiment of the song while using similar rhyming conventions and maintaining the cadence. Try to match the syllable count to the original verses.

###
[Paroles de "Love Story"]

[Couplet 1]
L'âme en peine
Il vit, mais parle à peine
Il attend
Devant cette photo d'antan
Il, il n'est pas fou
Il y croit c'est tout
Il la voit partout
Il l'attend debout
Une rose à la main
À part elle, il n'attend rien

[Couplet 2]
Rien autour n'a de sens
Et l'air est lourd
Le regard absent
Il est seul et lui parle souvent
Il, il n'est pas fou
Il l'aime c'est tout

Il la voit partout
Il l'attend debout
Debout une rose à la main
Non, non plus rien ne le retient

[Refrain]
Dans sa love story
Dans sa love story
Dans sa love story
Sa love story

[Couplet 3]
Prends ma main
Promets-moi que tout ira bien
Serre-moi fort
Près de toi, je rêve encore
Oui, oui, je veux rester
Mais je ne sais plus aimer
J'ai été trop bête
Je t'en prie, arrête
Arrête, comme je regrette
Non, je ne voulais pas tout ça

[Pre-refrain]
Je serai riche
Et je t'offrirai tout mon or
Et si tu t'en fiches
Je t'attendrai sur le port
Et si tu m'ignores
Je t'offrirai mon dernier souffle de vie

[Refrain]
Dans ma love story

Dans ma love story
Dans ma love story
Ma love story

[Couplet 4]
Une bougie
Peut illuminer la nuit
Un sourire
Peut bâtir tout un empire
Et il y a toi
Et il y a moi
Et personne n'y croit
Mais l'amour fait d'un fou un roi
Et si tu m'ignores
J'me battrais encore et encore

[Refrain]
C'est ta love story
C'est ta love story
C'est l'histoire d'une vie
Love story

[Outro]
Des cris de joie
Quelques larmes on s'en va
On vit dans cette love story
Love story, love story, love story

Now, I was personally very impressed by the results ChatGPT came up with. The translation rhymes, yet the word choice here retains the simplistic yet emotional tone of the original (Response 11-4).

Response 11-4. A beautifully translated version of Indila's "Love Story" in English

Here is an English adaptation of "Love Story" by Indila, maintaining the sentiment, rhyming, cadence, and similar syllable counts:

[Lyrics of "Love Story"]

[Verse 1]
A heart in pain,
He lives but speaks in vain.
He just waits,
Before this photo of old days.
No, no, he's not mad,
He believes, and that's all he has.
He sees her everywhere,
He's standing there,
A rose in his hand,
For her alone, he'll always stand.

[Verse 2]
Nothing near has meaning,
The air feels heavy, fleeting.
With an empty stare,
He's alone, yet speaks to her there.
No, no, he's not mad,
He loves her, that's all he has.
He sees her everywhere,
He's standing there,

Still with a rose in his hand.
No, nothing else can make him stand.

[Chorus]
In his love story,
In his love story,
In his love story,
His love story.

[Verse 3]
Take my hand,
Promise me we'll make it again.
Hold me tight,
Near to you, I'll dream all night.
Yes, yes, I want to stay,
But I don't know love's way.
I was such a fool,
Please stop, be cruel.
Stop now, I regret it all.
No, I never meant for us to fall.

[Pre-Chorus]
I'll be rich,
And I'll give you all my gold.
If you don't care,
I'll wait for you at the shore.
And if you ignore,
I'll give you my last breath of life.

[Chorus]

In my love story,
In my love story,
In my love story,
My love story.

[Verse 4]

A candle flame
Can brighten up the darkest plain.
A single smile
Can build an empire mile by mile.
And there is you,
And there is me,
No one believes,
But love turns a fool to royalty.
And if you ignore,
I'll fight for us forevermore.

[Chorus]

It's your love story,
It's your love story,
It's the tale of a life,
Love story.

[Outro]

Cheers and cries,
And tears fill our goodbye.

```
We live in this love story,
Love story, love story, love story.

---

This adaptation keeps the poetic rhythm while retaining the
emotional depth and imagery of the original French lyrics. Let
me know if you'd like any further refinements!
```

ChatGPT was able to successfully translate these beautiful song lyrics. As a polyglot (meaning a person that speaks several languages), I can personally attest to how well this translation adheres to the original meaning and intent of the song while maintaining the rhyming scheme very well. *C'est formidable.*

But that's not the only way we can alter media to suit our own purposes. Have you ever gifted a loved one with something personalized just for them? Things like mugs or throw pillows with a family member's face on them are great presents because the personal touch makes them sentimental.

So, let's look at how we can change a story to personalize it for a young child that would love to see a story tailored just for them.

Making the Tale of Goldilocks and the Three Bears More Personal

Telling children stories before bed is a beautiful way to bond, and there's nothing that excites them more than seeing themselves in the protagonist of the story. With ChatGPT, we can take these stories and make them catered to your own close loved ones. Try this out with your favorite bedtime story to tell your kids!

For this example, we're going to ask ChatGPT to tell us the story of Goldilocks and the Three Bears (Prompt 11-5).

Prompt 11-5. Getting ChatGPT to tell us a story

```
Tell me the story of Goldilocks and the Three Bears
```

In Response 11-5, you'll see how ChatGPT recounts the story for us.

Response 11-5. ChatGPT tells the story of Goldilocks and the Three Bears

```
The story of Goldilocks and the Three Bears is a classic fairy
tale with many versions, but here's a popular retelling:

---
```

Goldilocks and the Three Bears

```
Once upon a time, in a cozy little cottage in the forest,
there lived three bears: a big Papa Bear, a medium-sized Mama
Bear, and a tiny Baby Bear. Each bear had their own bowl for
porridge, their own chair, and their own bed.

One morning, the bears made porridge for breakfast, but it was
too hot to eat. They decided to go for a walk in the forest
while it cooled.

Nearby, a curious little girl named Goldilocks was wandering
through the woods. She stumbled upon the bears' cottage and,
finding the door open, decided to go inside.

---
```

The Porridge

```
In the kitchen, Goldilocks saw three bowls of porridge. She
tasted the first bowl.

"Too hot!" she exclaimed.

She tried the second bowl.
```

"Too cold!"

Finally, she tasted the porridge in the smallest bowl.

"Just right!" she said happily, and she ate it all up.

The Chairs
Feeling full, Goldilocks went into the living room and saw three chairs. She sat in the first chair.

"Too hard!" she said.

She tried the second chair.

"Too soft!"

Then she sat in the smallest chair.

"Just right!" she said. But as she sat back to relax, the little chair broke into pieces.

The Beds
Now feeling sleepy, Goldilocks went upstairs to the bedroom. She lay down on the first bed.

"Too hard!"

She tried the second bed.

"Too soft!"

Then she lay down on the smallest bed.

"Just right!" she sighed and fell fast asleep.

The Bears Return

Soon, the three bears came home from their walk. They noticed someone had been in their house.

"Someone's been eating my porridge!" growled Papa Bear.

"Someone's been eating my porridge!" said Mama Bear.

"Someone's been eating my porridge, and they ate it all up!" cried Baby Bear.

In the living room, they found the chairs.

"Someone's been sitting in my chair!" growled Papa Bear.

"Someone's been sitting in my chair!" said Mama Bear.

"Someone's been sitting in my chair, and they broke it!" cried Baby Bear.

When they went upstairs, they saw the beds.

"Someone's been sleeping in my bed!" growled Papa Bear.

"Someone's been sleeping in my bed!" said Mama Bear.

"Someone's been sleeping in my bed, and she's still there!" cried Baby Bear.

Goldilocks Wakes Up

Goldilocks woke up and saw the three bears staring at her. She screamed, jumped out of bed, and ran as fast as she could out of the house and into the forest. She never returned to the bears' cottage again.

And from that day on, Goldilocks learned to be more careful about wandering into houses that weren't hers.

The End.

Now let's ask ChatGPT to customize the story for us. I encourage you to try this for yourself by changing the protagonist, the bears, the setting, or anything else to make it more personal. In the meantime, check out how we did it in Prompt 11-6.

Prompt 11-6. Asking ChatGPT to give us a personalized version of Goldilocks and the Three Bears

I have two sons named Charlie and Emmett, and I want to tell them the story with the two of them as the protagonists. I want them to relate to the story on a personal level, so I want the story to take place in a small neighborhood in Oregon. Instead of bears, I'd like them to go into the house of several characters from Rick and Morty.

In Response 11-6, ChatGPT does an excellent job of integrating the changes we asked for, as well as adding flair based upon the *Rick and Morty* TV show.

Response 11-6. ChatGPT gives us a personalized story that's sure to delight

Here's a customized version of the story for Charlie and Emmett, set in a small Oregon neighborhood featuring characters from **Rick and Morty**:

Charlie and Emmett's Adventure

One sunny afternoon in a quiet neighborhood in Oregon, Charlie and Emmett decided to go exploring while their parents were busy gardening. The brothers loved adventure, and they never knew what they might discover.

As they wandered down a quiet street, they came upon a strange house at the end of the block. It looked like something out of a sci-fi movie, with gadgets sticking out of the roof and green glowing lights in the windows.

"Do you think anyone lives here?" asked Charlie.

"I don't know," said Emmett, "but we should check it out!"

The House of Rick and Morty

The boys knocked on the door, but no one answered. They pushed it open cautiously and stepped inside. The house was messy, filled with weird inventions and strange smells. Unbeknownst to them, the house belonged to Rick, Morty, and Summer, who had gone on an interdimensional adventure and left the door unlocked.

The Snacks

In the kitchen, Charlie and Emmett found three bowls of something that looked like glowing cereal.

Charlie tried the first bowl. "Too spicy!" he exclaimed, sticking out his tongue. It was clearly one of Rick's crazy experiments.

Emmett tried the second bowl. "Too sour!" he said, his face scrunching up.

Then they both tried the smallest bowl, which looked normal.

"Just right!" they said together, and they ate it all up.

The Chairs

Next, they went into the living room, which was cluttered with futuristic furniture. There were three chairs.

Charlie sat in the first chair. "Too stiff!" he said, standing up quickly.

Emmett tried the second chair. "Too wobbly!" he said as it started spinning uncontrollably.

Then they both tried the third chair, a cozy recliner in the corner. "Just right!" they said, curling up together. But when they pulled the lever, the chair launched them halfway across the room!

The Beds

After recovering from the chair fiasco, the boys found a staircase leading to the bedrooms. Upstairs, there were three beds.

Charlie flopped onto the first bed. "Too hard!" he said, rubbing his back.

Emmett tried the second bed, which was covered in glowing slime. "Too sticky!" he said, hopping off immediately.

Then they both crawled into the third bed, which was soft and smelled like freshly washed blankets.

"Just right!" they said, yawning. Before they knew it, they were fast asleep.

The Return of Rick, Morty, and Summer
Meanwhile, Rick, Morty, and Summer returned from their interdimensional trip, carrying strange gadgets and arguing loudly.

"Why is the door open?!" Rick growled.

As they walked into the house, they noticed the glowing cereal had been eaten. "Someone's been eating my toxic cereal!" Rick shouted.

"And someone sat in my hover chair!" complained Morty.

When they went upstairs, Summer stopped and pointed. "Uh, Rick? Someone's sleeping in my bed…and it's two kids!"

Charlie and Emmett woke up to find the three staring at them. Rick was holding a portal gun, and Morty looked panicked.

"Uh, hi?" said Charlie nervously.

"Kids, you've got about three seconds to explain yourselves before I send you to another dimension!" Rick barked.

A Quick Escape
Terrified, Charlie and Emmett scrambled out of bed and ran down the stairs. They darted past the strange gadgets, through the glowing kitchen, and out the front door.

"Let's never go in that house again!" said Emmett, panting as they ran home.

"Agreed!" said Charlie, laughing now that they were safe.

From that day on, Charlie and Emmett stuck to exploring places that didn't look like they belonged in a sci-fi cartoon. But they always remembered their wild adventure in the house at the end of the block.

The End.

What do you think? Would you like to tweak anything to make it even more personal?

As you can see from Response 11-6, ChatGPT comes up with a very entertaining story that still resembles the setup from Goldilocks and the Three Bears while adding the personal touch of making two young boys the protagonists. It also makes the story a joy to read for fans of *Rick and Morty*.

Conclusion

In this chapter, we looked at how ChatGPT can be used to rewrite and rephrase text for business, translation, and even the customization of a bedtime story to make it more personal. In the next chapter, we're going to be looking at how ChatGPT can be used to extract data for various uses, from conducting research based on blog reviews to creating a budget plan based upon actual buying habits.

CHAPTER 12

Budget Planning, Product Research, and Writing an Article with ChatGPT

At this point, we know that ChatGPT is good at handling large quantities of data—whether it be text, documents, or even photos. A good rule of thumb is that any time you're dealing with a lot of information, *it's the perfect opportunity to use ChatGPT*. ChatGPT performs the best with as much information as possible. Because of this, if you're conducting research on large quantities of information or are trying to sort through data for specific information, you know which tool to rely on.

In this chapter, we're looking at how ChatGPT can be used to extract data and do something with it. What do we mean by this? Well, for example, if you're conducting product research, it's unlikely you're just going to collect that information and then proceed to do nothing else. You're probably going to write up a report, create a marketing strategy, or use the information and write about it. If you're looking at financial information like bills and payment stubs, you're likely to then start making a budget, manage your spending habits, or categorize payments so that they're easier to sort through in the future.

L. Evelyn, *Making ChatGPT Work for You*, https://doi.org/10.1007/979-8-8688-1445-7_12

So, to save ourselves some time, we're going to use ChatGPT to reduce the effort needed to pour over large quantities of text and numbers by having ChatGPT follow two steps: analyze the material, then do something with it.

What will be covered in this chapter:

- Planning on writing a novel? Get to know your audience first! Conducting product research on fantasy novels.

- Get the most out of what you learned by sharing with others. Writing about writing based upon great examples.

- Take the guessing out of budget planning. Using ChatGPT o1 to take photos of actual receipts and make an effective budget plan to accurately cut spending costs.

Conducting Product Research Based on Blog Reviews

If you're selling a product of any kind, product research is crucial for understanding what interested consumers are looking for. You need to understand how other products may or may not be meeting their needs, what impresses your potential clients, and find out what's trending. Going through customer reviews, however, is extremely tedious.

Besides, not all reviews are the same. Imagine you're managing a cooking blog, and one of your readers leaves a bad review, but they rarely cook. Trying to impress that kind of reader isn't going to help your research in the long run if you're looking through your comments for feedback. You want people that enjoy cooking often to enjoy your blog, because they're the ones that will keep coming back. From a business perspective, the enthusiast's opinion matters more than someone that only dabs in cooking on occasion.

For this reason, in this example, we're going to be looking at how ChatGPT can be used to extract content from blog reviews that are directly related to the subject we're interested in. For this example, we're going to be examining reviews and critiques of popular fantasy novels. Blog reviewers are not only sure to be dedicated to the product they're discussing, they also have the ability to influence people interested in the topic they're discussing. Finding out about their opinion is much more efficient than, for example, just going through Amazon reviews of popular novels arbitrarily. So, let's get started.

Using ChatGPT to Compare Popular Fantasy Novels

In this example, we're going to examine the reviews of different fantasy novels and find out what reviewers agree are the common highlights and downfalls of each. We're going to use ChatGPT to find different bloggers to conduct our research and save time.

This example will consist of a back-and-forth conversation with ChatGPT that will use its web search abilities as well as reference previous points in the same conversation. The approach here will resemble concepts covered in Chapter 4, so if you want a reminder on how you can use ChatGPT to write content based upon information it was taught as reference, you can check that out. Now, let's get started with Prompt 12-1.

Prompt 12-1. Using ChatGPT to conduct product research for fantasy novels

I want you to find some book bloggers that have reviewed the book Game of Thrones, by George R. R. Martin. Give me the links to their blogs. Find bloggers that wrote negatively about the book, as well as bloggers that wrote positively about it. Give me a list. Make sure that the bloggers you find mostly read fantasy.

As you can see from our prompt, we're enlisting ChatGPT to find various sources to obtain information from bloggers that have reviewed *Game of Thrones* by George R. R. Martin. Not only are we trying to get their opinions, we're trying to specifically get the opinions that matter most to us—fans of fantasy. This will give us the viewpoint of an already-engaged audience in this genre. Response 12-1 shows the answer we got from ChatGPT.

Response 12-1. ChatGPT finds various book reviews of various opinions on Game of Thrones, by George R. R. Martin

Here is a list of book bloggers who have reviewed *A Game of Thrones* by George R.R. Martin, with both negative and positive perspectives, focusing on bloggers who primarily review fantasy literature:

Negative Reviews:
1. **Kelson Vibber** - On the blog *Hyperborea*, Kelson Vibber found *A Game of Thrones* difficult to get into, describing it as a "tedious" read with characters he struggled to care about. He compared it unfavorably to other epic fantasies like *The Wheel of Time*.

2. **John Daulton** - On his blog *John Daulton Books*, Daulton criticized the book for its unnecessary sexual content and inconsistent dialogue, though he did praise its descriptive passages and character depth.

3. **Fantasy Book Review** - A reviewer on *Fantasy Book Review* expressed frustration with the character development, stating that the emphasis on morally gray characters made it hard to care about them, ultimately finding the book boring despite its creative world-building.

Positive Reviews:

1. **Books of Brilliance** - *Books of Brilliance* praised *A Game of Thrones* as a must-read for fantasy lovers, emphasizing its detailed world-building and its importance in redefining the fantasy genre.

2. **Tommye** - On *The Fantasy Review*, Tommye praised *A Game of Thrones* for its complex politics and intricate character relationships, ranking it as one of the top epic fantasies.

3. **Fantasy Book Review** - Another review on *Fantasy Book Review* from a different reviewer lauded the novel as one of the greatest fantasy works, highlighting its suspenseful narrative and emotional depth.

These bloggers provide a balanced view of the book, with some focusing on its flaws and others on its strengths, all within the context of the fantasy genre.

In Response 12-1, ChatGPT provides a variety of sources to draw from in our search for critical opinions on *Game of Thrones*. Now, you'll see in Prompt 12-2 that we're continuing our conversation with ChatGPT and asking it to follow the same procedure for another book, this one entitled *Sword Catcher* by Cassandra Clare.

Prompt 12-2. Asking ChatGPT to do more research

Now do the same for the Sword Catcher by Cassandra Clare

Our prompt is very simple because we're using our earlier prompt as a guide. You'll see in Response 12-2 that ChatGPT easily understands what we're asking for and complies.

Response 12-2. ChatGPT gives a list of bloggers and reviews for Cassandra Clare's Sword Catcher

Here are some book bloggers who have reviewed *Sword Catcher* by Cassandra Clare, with both negative and positive perspectives, focusing on those who typically review fantasy literature:

Negative Reviews:
1. **GeekDad** - While acknowledging the strong world-building in *Sword Catcher*, the reviewer found the book to be somewhat slow-paced, with much of the narrative feeling like setup for future installments. The feeling of being left hanging after so many pages was a drawback, although the depth of the characters and setting did leave the reviewer interested in what comes next.

2. **Before We Go Blog** - This review highlighted some clunky aspects of Clare's use of languages and cultural inspirations, which occasionally took the reviewer out of the story. Additionally, the in-world mythos interludes were seen as redundant, though the book's strong character development and intriguing plot were praised.

Positive Reviews:
1. **WinterIsComing.net** - This blog praised *Sword Catcher* as possibly Cassandra Clare's best work yet, appreciating its refreshing departure from her previous YA series. The detailed

world-building, complex character relationships, and the
novel's ability to surprise even seasoned Clare readers were
highlighted as major strengths.

2. **Fantasy Book Critic** - The reviewer commended the book for
its solid blend of character-driven and plot-driven elements.
The evolving relationships between characters like Kel and Lin,
as well as the intrigue surrounding the political machinations
in the city of Castellane, were particularly well-received.

These bloggers provide a mix of perspectives on *Sword Catcher*,
offering insights into both its strengths and potential
weaknesses within the fantasy genre.

We have more blogs and reviews to examine! We're going to follow the
same steps and get bloggers and reviews for one more book, this time for
The Ruin of Kings by Jenn Lyons (Prompt 12-3).

Prompt 12-3. A final query for further research

Now do the same for The Ruin of Kings by Jenn Lyons

In Response 12-3, you'll see ChatGPT once again adheres to what was
requested in our short prompt.

Response 12-3. ChatGPT finds diverse reviews for The Ruin of Kings
by Jenn Lyons

Here are some book bloggers who have reviewed *The Ruin of Kings*
by Jenn Lyons, with both negative and positive perspectives,
focusing on those who primarily review fantasy literature:

Negative Reviews:
1. **Novel Notions** - The reviewer found *The Ruin of Kings*
overly complex and felt that the narrative structure, which

jumps between timelines and perspectives, was unnecessarily convoluted. They appreciated the world-building and side characters but were not as invested in the main character, Kihrin, which affected their overall enjoyment of the book.

2. **Fantasy Book Review** - The reviewer noted that while parts of the book were exceptional, the dual narrative format and the complexity of the plot could be confusing. They also mentioned that the ending felt somewhat rushed and overwhelming, making it difficult to fully grasp the story's conclusion.

Positive Reviews:

1. **Dragonmount** - This blog praised *The Ruin of Kings* for its intricate world-building and complex characters. The reviewer appreciated the depth of the story and the way it tackled themes like fate and destiny. They found the novel engaging and are eager to read the next installment in the series.

2. **Reading Reality** - The reviewer highlighted the rich, well-developed world and the compelling narrative. They enjoyed the complex plot and the layers of mythology that Jenn Lyons incorporated into the story, finding it a rewarding read for fans of epic fantasy.

These reviews provide a balanced view of *The Ruin of Kings*, offering insights into both its strengths and potential weaknesses within the fantasy genre.

Perfect. Now that we have gathered reviews and bloggers covering several fantasy novels, let's put all this information to use.

Summarizing Information Based upon Product Research

Now that we have conducted a little product research, let's ask ChatGPT to answer two questions: What were the most common complaints by these avid readers, and what did they enjoy the most about these books? We're using Prompt 12-4 to find out.

Prompt 12-4. Using ChatGPT to conclude our research

Based upon all the books we've examined, what would you say is most important to readers of these books and genres? What are they looking for, and what are they hoping to avoid?

ChatGPT's answer can be seen in Response 12-4.

Response 12-4. ChatGPT summarizes important information from its research

Based on the reviews and analysis of *A Game of Thrones* by George R.R. Martin, *Sword Catcher* by Cassandra Clare, and *The Ruin of Kings* by Jenn Lyons, several key factors stand out as being particularly important to readers of epic and high fantasy. Here's what they generally look for and hope to avoid:

What Readers Are Looking For:

1. **Complex World-Building:**
   - Readers of these genres highly value intricate and well-developed worlds. They appreciate settings that feel immersive and real, with rich histories, detailed political systems, and diverse cultures. For example, *The Ruin of Kings* was praised for its "layered and intricate" world-building, while *Sword Catcher* was noted for its "seamless diversity" and multicultural influences.

2. **Engaging Characters:**

- Well-drawn, multifaceted characters are crucial. Readers enjoy characters who are complex, flawed, and relatable. Protagonists with depth, as well as well-crafted side characters, can significantly enhance a reader's connection to the story. *The Ruin of Kings* received praise for its memorable characters and the engaging banter between them.

3. **Intriguing Plot and Narrative Structure:**

- An engaging, well-paced plot is essential, but readers also appreciate creative narrative structures that add depth to the story. However, there is a fine line; if the narrative becomes too convoluted, it can detract from the experience. *A Game of Thrones* was lauded for its political intrigue and complex character relationships, while *The Ruin of Kings* was both praised and criticized for its dual timelines.

4. **Themes of Power and Morality:**

- Themes like power, fate, and morality resonate deeply with fantasy readers. They are drawn to stories that explore these concepts in nuanced ways, often through the lens of political intrigue or moral dilemmas. The exploration of these themes in *The Ruin of Kings* was particularly noted for making readers reflect on the nature of fate and control.

What Readers Hope to Avoid:

1. **Overly Convoluted Storytelling:**

- While readers appreciate complexity, they can be put off by storytelling that feels overly complicated or difficult to follow. This was a common criticism of *The Ruin of Kings*, where some found the narrative structure confusing and hard to engage with.

2. **Slow Pacing:**
   - Readers often express frustration with books that have slow pacing, particularly in the opening sections. If the plot takes too long to develop or feels bogged down by excessive detail, it can lead to disengagement. For instance, some reviewers of Sword Catcher found the pacing slow, especially in a book with as many pages as it had.

3. **Unnecessary Complexity or Jargon:**
   - While detailed world-building is appreciated, readers do not favor excessive use of made-up languages or complex jargon that can interrupt the flow of reading. This was a minor critique in Sword Catcher, where some felt the use of multiple languages could be distracting.

4. **Dark or Overly Grim Tones:**
   - Although many readers are drawn to the darker aspects of epic fantasy, such as themes of war, betrayal, and moral ambiguity, there's a balance to be struck. When these elements are overemphasized without moments of relief or hope, it can become overwhelming. A Game of Thrones is often discussed in this context, where the relentless grimness can be a turn-off for some.

In summary, readers of these genres seek a balance of rich, immersive world-building, complex and relatable characters, and a plot that is engaging without being overly convoluted. They appreciate depth and complexity but are wary of anything that detracts from the clarity and enjoyment of the story.

We've given ourselves an excellent summary of what are essentially customer reviews of very popular fantasy novels, which can give us a guide to follow when writing something on our own. This is an excellent summary for product research. If you wanted, you could go deeper into this analysis, look into the original blogs ChatGPT is referencing, and ask more questions about keywords these novels use or the audiences that resonated the most with these novels.

But here's an idea. We have a lot of valuable insights to work with for ourselves. Wouldn't other writers be invested in understanding these insights as well? This would be a perfect opportunity to take something we've learned and use it for yet another purpose by writing a blog post we can share with like-minded individuals and grow a community for marketing purposes. So, let's take this a step further and ask ChatGPT to write a guide to writing like the greats.

Creating a Guide to Writing Fantasy Inspired by the Greats

In our next prompt, we're going to ask ChatGPT to take all the information we've gathered thus far—bloggers, links, and critical reviews—and write a guide to writing fantasy for beginners. Not only is this a great resource for us, we can turn it into interesting content for fellow writers looking to improve their craft. In our prompt, we've also given ChatGPT specific instructions on the tone to use in its writing to ensure it has a decent readability and doesn't sound like a standard response from ChatGPT that can sound disingenuous. This is exactly what we did in Chapter 5, which is where you can find more information on how to prompt ChatGPT to adjust its writing tone.

Prompt 12-5. Using ChatGPT to turn the results of our research into
an article

Now, based upon all of the information you've gathered
from various book bloggers, I want you to write a guide to
writing high fantasy for beginners. Write it in prose, in a
conversational tone. Using these books and reviews as points of
reference, explain to readers how they can avoid pitfalls that
can make their writing weaker. I also want you to detail how
aspiring authors should take these great works of fiction as
inspiration by referencing the strongest points in the critical
reviews we've examined thus far.

As you can see, in Response 12-5, ChatGPT writes an excellent guide
for aspiring writers and veterans alike.

Response 12-5. ChatGPT writes the ultimate guide to writing
fantasy based upon the greats

A Beginner's Guide to Writing High Fantasy: Lessons from Great Works

Writing high fantasy is like stepping into a vast, untamed
wilderness of your own creation. It's a genre that allows for
the deepest exploration of human nature, the most intricate of
worlds, and the grandest of tales. But with such freedom comes
the responsibility to craft a story that captivates, resonates,
and remains accessible to your readers. To help you on your
journey, here's a guide based on the lessons learned from some
of the genre's most notable works and the critiques they've
received from seasoned readers and reviewers.

1. Build an Immersive World, But Beware of Overloading

World-building is the cornerstone of high fantasy. Your readers want to be transported to a place that feels as real as their own, yet filled with magic, dragons, and ancient prophecies.

Lesson from *The Ruin of Kings* by Jenn Lyons:

Reviewers have praised Lyons for her rich, layered world-building. However, some also felt that the complexity of her world sometimes made the narrative difficult to follow. The key takeaway here is to ensure your world-building enhances the story rather than overwhelms it. Introduce your world gradually, weaving in details as they become relevant to the plot, rather than bombarding your readers with information all at once. This will keep them engaged without feeling lost.

2. Craft Complex Characters with Relatable Motivations

In high fantasy, your characters are often the lens through which readers experience your world. They should be as complex and multifaceted as the world around them.

Lesson from *A Game of Thrones* by George R.R. Martin:

Martin's characters are renowned for their depth and moral ambiguity. This complexity keeps readers invested because they can see shades of themselves in the characters' struggles. However, it's important to balance this with characters that readers can still root for. While moral grayness is compelling, your characters need to have clear, relatable motivations that drive the story forward.

3. Engage with Intriguing Plotlines, But Keep It Coherent

An intricate plot with twists and turns is a hallmark of high fantasy. Your readers should feel like they're piecing together a grand puzzle, with each revelation bringing them closer to the truth.

Lesson from *Sword Catcher* by Cassandra Clare:

Clare's *Sword Catcher* was praised for its intricate plot and
the relationships between characters, which kept readers
hooked. However, some reviewers noted that the pacing could
be slow, especially in such a long book. This highlights the
importance of maintaining a balance between complexity and
coherence. Your plot should be engaging, but it should also be
clear enough that readers can follow without getting lost. If
you have multiple storylines, ensure they converge in a way
that feels natural and satisfying.

4. Use Unique Narrative Structures, But Don't Overcomplicate

High fantasy often experiments with narrative structures,
such as dual timelines or multiple points of view. These can
add depth and variety to your storytelling, but they can also
confuse readers if not handled carefully.

Lesson from *The Ruin of Kings* by Jenn Lyons:

The dual narrative and non-linear timeline in *The Ruin of Kings*
were both praised for their creativity and criticized for their
complexity. If you choose to use such structures, be sure they
serve the story. Consider how each narrative choice helps to
reveal character, develop the plot, or enhance the theme.
And always keep your readers in mind—clarity should never be
sacrificed for the sake of experimentation.

5. Tackle Dark Themes with Care

Themes like power, corruption, and moral ambiguity are staples
of high fantasy. These elements add gravitas to your story,
making it resonate on a deeper level.

Lesson from *A Game of Thrones* by George R.R. Martin:
Martin's exploration of dark themes like betrayal, power
struggles, and the consequences of honor have become
iconic. However, some readers find the relentless grimness
overwhelming. It's important to balance dark themes with
moments of hope or levity. This not only makes your story more
palatable but also highlights the darkness by contrast.

6. Engage with Readers' Expectations, Then Subvert Them

High fantasy readers often come with certain expectations,
whether it's the hero's journey, epic battles, or magical
quests. Playing with these tropes can make your story feel
fresh and exciting.

Lesson from *Sword Catcher* by Cassandra Clare:
Clare's work was noted for subverting the typical expectations
of fantasy romance, offering something different that intrigued
readers. Consider how you can take familiar elements of the
genre and twist them in unexpected ways. This keeps your story
engaging and helps it stand out in a crowded field.

7. Pacing: Maintain Momentum, Avoid Drag

High fantasy can span thousands of pages and multiple volumes,
so maintaining a good pace is crucial. Readers should feel like
the story is always moving forward, even during quieter moments.

Lesson from *The Ruin of Kings* and *Sword Catcher*:
Both books received feedback about pacing—*The Ruin of Kings*
was noted for its thrilling middle section but had a slower
beginning and a somewhat rushed ending, while *Sword Catcher* was
critiqued for slow pacing in parts. Aim to keep your narrative
rhythm consistent. Use slower moments to build tension, develop
characters, or deepen the world, but ensure these moments serve
the story and lead to something significant.

Conclusion: Crafting a Compelling High Fantasy

Writing high fantasy is a challenging but rewarding endeavor. By focusing on immersive world-building, complex characters, and engaging plots while avoiding common pitfalls like excessive complexity and uneven pacing, you can craft a story that resonates with readers. Take inspiration from the works of Martin, Clare, and Lyons—learn from their strengths and understand where they stumbled. Remember, your goal is to create a world that feels real, characters that readers care about, and a story that keeps them turning the pages.

Most importantly, have fun with it. High fantasy offers a playground for your imagination—so let your creativity soar, but keep your readers in mind as you do. Happy writing!

This short article ChatGPT has written is fantastic for informing authors on ways they can improve their writing. And with the information we've given ChatGPT both through its own research inspired by our prompts, we could continue to create more content. We could go on to create more guides, reports, presentations, and more.

Now we're going to look at how we can give ChatGPT information that's even harder to sift through: numbers. We're going to use ChatGPT to create a budget plan based upon actual spending habits from pictures of receipts.

Using ChatGPT o1 to Make a Budget Plan

Budgeting is hard because the most you can ever do is work with estimates when you're trying to examine your spending habits—that is, unless you're willing to pore over every receipt you've ever gotten and break out the calculator.

But as we know, ChatGPT can analyze images! This means that we can send ChatGPT an image of our receipt and let it do the rest. So, we're going to do just that. We're going to ask ChatGPT to tell us about our spending habits by sending photos of actual receipts. Not only will ChatGPT be able to read and itemize our spending habits based upon factual data, it will then be able to tell us about where we're spending the most and how we can reduce costs.

But let's answer the question that I'm sure is on your mind. Why are we using ChatGPT o1 this time? First, let's talk about what ChatGPT o1 is and what it does.

What Is ChatGPT o1? What Is It Used For?

ChatGPT o1 is the model that takes longer to *think* about its answer before providing it. What this means is that rather than prioritizing providing a quick answer, it prioritizes an answer that accurately completes your request. "Under the hood," it takes more processing power and therefore isn't as appealing if you're looking for a quick answer. Also, because it isn't as cost effective, it also tends to provide shorter answers than other ChatGPT models.

But, because this functionality takes a longer amount of time coming up with a precise answer, it's more useful for our purpose—namely, crunching numbers and calculating savings based upon photos of receipts. We want accurate numbers and calculations, and we really want ChatGPT to take its time analyzing the text from the photo so that there aren't any errors in interpreting the information.

In Figure 12-1, you'll see that the ChatGPT o1 model actually tells you how long it spent "thinking."

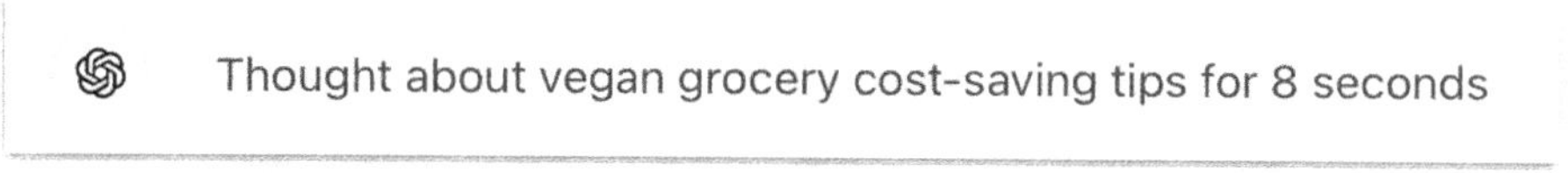

 Thought about vegan grocery cost-saving tips for 8 seconds

Figure 12-1. *The ChatGPT o1 model tells you how long it spent thinking about its answer. For some reason, I find this funny*

With that in mind, let's move on to our example. Remember to switch over to ChatGPT o1 for this example if you're following along!

Giving ChatGPT Receipts to Analyze for Budget Planning

To begin with, we're going to upload a photo of a grocery receipt that we want ChatGPT to analyze. We'll be giving ChatGPT a receipt from an organic grocery store so we can get an idea of how we can save money based upon common spending habits. Figure 12-2 shows the image of the receipt we used for our prompt.

Total:	USD$ 96.06	
GROCERY		
CRISPY COCOA RICE	5.49	F
Voided Item		
CRISPY COCOA RICE	-5.49	F
EXTRA CREAMY OATMILK	3.99	F
ITALIAN SODA-ORG-POM	3.99	T F
*CRV FS/TX 10	0.10	T
ORG CHICKEN BROTH	4.99	F
ORG GREEN TEA	5.99	F
ORG PURE FRUIT SNACK	5.00	F
2 @ 2 FOR 5.00		
ORGANIC BLACK TEA	5.99	F
Voided Item		
ORGANIC BLACK TEA	-5.99	F
ORGANIC PUMPKIN	3.29	F
PASTA ORZO	1.67	F
1 @ 3 FOR 5.00		
XTRA VRGN OLIVE OIL	10.99	F
DELI		
CHEDDAR MAC&CHEESE	7.39	F
GRLC RSTD POTATOES	4.87	F
TEMPURA SHRIMP ROLL	10.99	F
BULK		
SUNFLOWER SEEDS	2.99	F
MEAT		
ORG 85% GROUND BEEF	8.99	F
POULTRY		
ORG CHICKEN THIGH	7.27	F
BAKERY		
TWO BITE BROWNIES	6.99	F
TWO BITE DESSERT	4.89	F
BOOKS		
AMETHYST BRACELET	7.99	T
ZORBITZ BRACELET	8.99	T
FROZEN		
CHICKEN POTSTICKERS	4.99	F
STAR CARD	-5.00	

Figure 12-2. *A receipt from a natural food grocery store*

Now that we've uploaded our photo, let's look at Prompt 12-6.

Prompt 12-6. Prompting ChatGPT to help us save money on groceries

This receipt represents what I spent on groceries on my last trip. I'm trying to save money, and I'd like to find a way to spend at least 10$ less on my next shopping trip. Help me figure out where I can cut costs.

The answer we got from ChatGPT can be seen in Response 12-6.

Response 12-6. ChatGPT calculates where we can save on grocery bills

Here's an analysis of your receipt and suggestions to help you save at least $10 on your next trip:

Higher-Cost Non-Essential Items
1. Amethyst Bracelet – $7.99
2. Zorbitz Bracelet – $8.99
   • Suggestion: Skip these purchases next time, saving $16.98. Jewelry items aren't essential for groceries.

Snacks & Treats
1. Two Bite Brownies – $6.99
2. Two Bite Dessert – $4.99
   • Suggestion: Limit to one dessert item or skip altogether, saving at least $6.99.

Pre-Made Deli Items

1. Cheddar Mac & Cheese - $7.39
2. Garlic Roasted Potatoes - $4.87
3. Tempura Shrimp Roll - $10.99
   - Suggestion: Consider making these at home or reducing deli purchases. Skipping the shrimp roll could save $10.99.

____

Potential Savings Summary
- Skipping jewelry: $16.98
- Reducing desserts: $6.99
- Cutting one deli item (shrimp roll): $10.99

Total Potential Savings: $34.96 (well over your $10 goal)

____

Other Ideas for Reducing Costs
- Organic Products: Consider swapping some organic items for non-organic. For example, non-organic chicken broth could be cheaper.
- Snacks: Reduce or skip packaged snacks, like the organic pure fruit snack ($5.00 for two).
- Buy in Bulk: Items like pasta and sunflower seeds can be bought in bulk for long-term savings.
- Compare Prices: Check if local stores have lower prices for staples like tea, meat, and pasta.

Would you like more suggestions, like meal planning strategies to save even more?

As you can see, we've been able to calculate savings that exceed our goal of saving $10! We even get some saving tips for future shopping trips, like where to buy products at the best prices.

Additional Tips on Budgeting with ChatGPT

From here, you can ask ChatGPT any number of questions about your spending habits. Obviously, we wouldn't show here any actual spending details like bank statements, credit card statements, or the like. But for your personal use, you can make ChatGPT even more useful for calculating general saving costs with even more accuracy by providing these documents for analysis. Just make sure sensitive information is blacked out!

Conclusion

In this chapter, we went over the different ways that ChatGPT can be used to extract information online, inform future decisions, create a blog post from collected information, make a budget planner, and get insights on future purchases. In the next chapter, we're going to finalize everything by having fun generating images and videos with ChatGPT and Sora.

CHAPTER 13

Visualize Your Ideas with ChatGPT's DALL-E and Sora

At this point, we've covered all of the amazing features included in ChatGPT except for one of the most exciting—image and video generation. This chapter is going to be fun. We're going to use ChatGPT's DALL-E's model to help us visualize our ideas by creating beautiful artwork and almost eerily realistic videos with the help of ChatGPT's Sora model.

Now, there's a lot of controversy surrounding the use of AI-generated art. In this book, we're not encouraging people to stop hiring artists. Rather, we're trying to encourage people to help themselves visualize the concepts they would otherwise struggle to imagine. We feel it's best to use these tools to better collaborate with team members—artists included— and find inspiration. It should be our responsibility to not replace human beings, but use tools that better help us work together. That is what we'll be covering in this chapter.

Likewise, we're going to cover how we can use the short videos we can create with Sora to make content that is **not** meant to **impersonate** or **deceive** by any means. Rather, we're going to create videos that can make content creation look more professional by adding short clips to make YouTube videos more engaging or adding movement to a website's landing page.

© Lydia Evelyn 2025
L. Evelyn, *Making ChatGPT Work for You*, https://doi.org/10.1007/979-8-8688-1445-7_13

So let's get started!

What's covered in this chapter:

- How do we use ChatGPT to create images anyway? Looking at how to use ChatGPT to generate images.

- When you can't find the words, paint a picture. Seeing how we can use ChatGPT's image generation capabilities to collaborate with team members for work.

- Get ideas and create assets for your own personal projects. Conceptualizing ideas for creative projects.

- Go crazy! Having fun generating cool art.

- Sora is completely different from everything we've done so far! Let's find out how to use it.

- Lights, camera, action! Direct Sora to make a smooth, lifelike video for websites and content creation.

- Being responsible with AI-generated content—the dos and don'ts of using AI-generated images and videos.

How Do We Use ChatGPT to Generate Images?

Well, it's actually quite simple. To get ChatGPT to create images for us, we're going to do the exact same thing we've been doing this whole time. We just ask. Let's start with Prompt 13-1.

Prompt 13-1. Using ChatGPT to generate images for the first time

```
Create an image of an apple
```

Notice that all we had to do was say "create an image." You could also phrase this slightly differently and say "make a photograph" or "generate artwork." Regardless of how you say it, you only need to ask ChatGPT in some way within the prompt to create this image for you.

The image ChatGPT generated as a result can be seen in Figure 13-1.

Figure 13-1. *ChatGPT creates the image of a nice, ripe-looking apple. Disclaimer: Figure 13-1 is an AI-generated image*

Would you just look at the detail! There are even a few droplets of moisture on the skin of the apple in this photo. Already this is quite impressive, but what if this isn't quite what we were looking for? We got an apple, but it's sitting in a pretty plain setting. Why don't we ask ChatGPT to try again, with a slight alteration? Like before, we're just going to ask for the change in Prompt 13-2.

Prompt 13-2. Telling ChatGPT to make a change to the image it created

```
Actually, I want the apple to be on a wooden table.
```

Now that we've asked for our adjustment, you'll notice that DALL-E generates a whole new image for us and abides by our request (Figure 13-2).

Figure 13-2. _ChatGPT creates a new image, this time with the apple on a table._
Disclaimer: Figure 13-2 is an AI-generated image

So now that we know how to get ChatGPT to generate images, let's have some fun. ChatGPT can not only create images, but it can also create art in different styles, create realistic-looking photos of different scenarios, and more. Let's get started!

Using ChatGPT to Generate Images for Work

In the workplace, even when you have a graphic artist on the team, creating graphic art is often a constant back-and-forth discussion as everyone tries to get on the same page creating the visuals. With ChatGPT, non-artistic people can get their point across by showing a workable example of what they want or even get the job done themselves by making a few adjustments to a response from ChatGPT.

For scenarios like these, you can treat ChatGPT like an extra coworker that can get drafts done quickly. And like a coworker, ChatGPT works best when you give it the whole story, with as many details as possible. Let's look at different scenarios for using ChatGPT to help with graphic design projects.

Advertising a Cleaning Product

In this scenario, a graphic designer needs to create a piece for a company that produces cleaning products. We need to tell ChatGPT what we're creating and the look we're trying to achieve in the final product (Prompt 13-3).

Prompt 13-3. Telling ChatGPT to create graphic art for advertising

```
I'm creating graphic art for a brand that makes cleaning
products. I need an illustration of a mop sweeping across the
floor of a kitchen, leaving a trail of sparkles.
```

Figure 13-3 shows the image ChatGPT generated in its response.

Figure 13-3. *A graphic for advertising a mop.*
Disclaimer: Figure 13-3 is an AI-generated image

As you can see in Figure 13-3, ChatGPT has given us a great starting point, but personally, I don't quite like the fact that the mop is an illustration, but the kitchen behind it is in 3D. Now, in order to correct this, I'm actually not going to give ChatGPT a lengthy description of what to or not to do. Sometimes you just need a refresh. Check out what I mean in Prompt 13-4.

Prompt 13-4. Sometimes getting different results from ChatGPT is very simple

```
Not quite. Try again
```

ChatGPT's fresh take on our request can be seen in Figure 13-4.

Figure 13-4. *ChatGPT creates a much more cohesive image for our graphic image.*
Disclaimer: Figure 13-4 is an AI-generated image

In my opinion, Figure 13-4 much more closely resembles what we'd be looking for in a graphic advertising cleaning products. The style is consistent because the kitchen is 3D, but the mop now looks much more realistic.

Drafting Up a Prototype

When you're collaborating with other people, it helps to get everyone on the same page with a visual. For example, imagine a production team needs an image to work with to create a prototype product. But the people that know how the product is supposed to work aren't usually the same people that know what it should look like. And sometimes, if it feels like you've been banging your head against a wall, you just need new ideas. So let's look at how we can prompt ChatGPT to create concept art for a smart watch prototype (Prompt 13-5).

Prompt 13-5. Instructing ChatGPT to create concept art for a smart watch

```
I'm making a promotion for a tech company making a new smart
watch with a simplistic design. We need some concept art to
hand to the team responsible for designing the prototype.
Generate a few images to give us ideas for how the minimalistic
smart watch might look like.
```

Figure 13-5 shows the image ChatGPT generated in its response.

Figure 13-5. *ChatGPT creates concept art for a minimalistic smart watch design.*
Disclaimer: Figure 13-5 is an AI-generated image

This is fantastic! ChatGPT has created exactly the kind of image we wanted. But let's say the team is having a hard time visualizing what the prototype would look like in different colors while keeping a minimalistic look. We're going to ask ChatGPT for changes in Prompt 13-6.

Prompt 13-6. Telling ChatGPT to give us options in different colors

This is great! Now give us options in different muted colors

Figure 13-6 shows the image ChatGPT generated in its response.

Figure 13-6. *DALL-E offers different color options for our smart watch... but something isn't right.*
Disclaimer: Figure 13-6 is an AI-generated image

In Figure 13-6, we got what we were looking for in that we were offered other color variations for our smart watch prototype, but they're cut out of focus, and it's hard to really see what they look like. Let's ask ChatGPT to make another adjustment, which you can see in Prompt 13-7.

Prompt 13-7. Making sure we can see other options properly

```
Make sure the different color options are easy to see
```

Figure 13-7 shows the image ChatGPT generated in its response.

Figure 13-7. *DALL-E expands on the image it created to allow us to see the results better.*
Disclaimer: Figure 13-7 is an AI-generated image

This worked out well! As you can see, using ChatGPT to create images isn't so different from how we've been using it thus far. It requires a little back-and-forth at times to get what you want, but when you use the right prompts, you can get results you can be happy with.

Creating a Hero Image for an Article or Blog Post

If you're the writer for a personal blog or an editorial, chances are you don't care too much about the hero image as long as it looks good. But most of the time, the images most relevant to what you're creating are copyrighted stock images that you need to pay for, and the prices can get surprisingly steep.

In this example, we're creating a hero image for a blog post about upcycling t-shirts to reduce waste. Thankfully, ChatGPT can make this process pretty simple. We instruct ChatGPT to create this image in Prompt 13-8.

Prompt 13-8. Creating a hero image for a blog post

```
I need you to make the hero image for a blog post I'm writing
on how to up-cycle old t-shirts for various DIY projects. I
need an image of a bunch of laid out t-shirts. It should look
like a stock image photo.
```

In Figure 13-8, you'll see that ChatGPT got it right the first time!

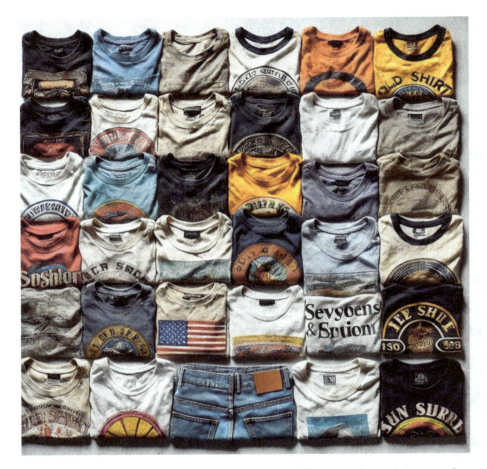

Figure 13-8. *ChatGPT generates a very well-created stock image for our blog post.*
Disclaimer: Figure 13-8 is an AI-generated image

This is a great hero image for a blog that perfectly expresses the concept we gave ChatGPT to work with.

Now let's look at more use cases that can be used for more creatively expressive purposes, like creating art for book covers and promotional illustrations.

Using ChatGPT to Help Visualize Concepts for Book Promotion

Speaking from experience, I can tell you that rarely ever is a good author also a good designer. More than that, while an author isn't usually a designer, they *very often* want their ideas made manifest in art. Nothing is more exciting than imagining an epic scene and then seeing it actually "happen" in a piece of artwork. Likewise, when it comes to other things like promoting the book, authors often find themselves at a loss when it comes time to create promotional graphics, especially if they're working on their own.

For this reason, ChatGPT is fantastic for visualizing ideas and take the author one step closer to actually seeing the radical ideas that play out in their minds. So let's make some creative artwork for a fantasy book.

Visualizing a Scene from a Steampunk Fantasy Book

Steampunk is one of those niche genres that is so uniquely distinct, yet hard to come across too often. Writers and readers of the genre alike love any iteration of art that depicts steampunk because of how hard it can be to find sources of inspiration. For this example, we're going to have ChatGPT illustrate an epic scene for a steampunk fantasy novel (Prompt 13-9).

Prompt 13-9. Asking ChatGPT to illustrate a scene from a book

```
I'm writing a steampunk fantasy novel and I want to visualize
the scene where a big blimp crashes into the Eiffel tower.
Create an image depicting this in the realism art style, at
night, with an explosion as the blimp crashes into the tower.
```

Figure 13-9 shows the image ChatGPT generated in its response.

Figure 13-9. *DALL-E creates a fantastic image of a blimp crashing into the Eiffel tower.*
Disclaimer: Figure 13-9 is an AI-generated image

The image DALL-E has given us is already fantastic. The explosion, in particular, looks incredible. And the blimp itself also looks like it fits the genre perfectly. But what if we prefer a more illustrated look? Something that looks a little less like a movie poster and more like a painting. We ask ChatGPT to accomplish this in Prompt 13-10.

Prompt 13-10. Don't be afraid to change your mind and ask for alterations on the fly

```
Actually, now I want the art to look more like an illustration
than an image
```

Figure 13-10 shows the image ChatGPT generated in its response.

Figure 13-10. *DALL-E recreates our epic scene in a beautiful illustration.*
Disclaimer: Figure 13-10 is an AI-generated image

351

Truly, either option looks fantastic. This just goes to show that some art is up to the viewer's perspective, but ChatGPT offers the ability to explore all the options.

Conceptualizing a Book Cover with DALL-E

For authors, traditionally published and self-published alike, getting cover art can be the most exciting part of publishing. Working with the cover artist is a thrilling process, but when an author has a specific idea in mind that the artist isn't understanding, the situation goes from exciting to frustrating. Even traditionally published authors, who will have little to do with the final cover art, are still asked for their opinion on how the end product will look like.

Again, ChatGPT offers itself as a wonderful means of communicating a vision within individuals trying to collaborate. In this example, we're going to look at how DALL-E can help an author visualize a cover for their romance book (Prompt 13-11).

Prompt 13-11. Conceptualizing a book cover for a romance novel

I'm writing a romance book, and I want inspiration for what the cover will look like so I can give the actual cover artist an idea of what I'm looking for. Create an image with the dimensions of a book cover of a woman and a man from the 1800s holding each other, with looks on their faces as though they know they shouldn't be doing what they're about to do. The image should show them from the waist up. The color scheme should be rich and warm. This is a sensual but heart-warming story. Generate the image

Figure 13-11 shows the image ChatGPT generated in its response.

Figure 13-11. *A perfect mock-up of a romance book cover that romance readers would fawn over.*
Disclaimer: Figure 13-11 is an AI-generated image

Creating Promotional Graphics for a Book Cover Reveal

If you read a lot of fiction and follow fiction authors on social media, you know that an exciting new release isn't complete without a cover reveal. This is when the author shows the audience what the beautiful book cover will look like ahead of release, and in some cases, the cover release is so successful that readers will buy out the preorders and the bookstores when the book goes on sale officially.

However, for all of this to go as planned, one has to prepare the social media posts that will advertise the book to an excited audience. For this reason, authors will often provide a backdrop to the cover art they're presenting for their audience. For this example, we're going to ask ChatGPT to create the backdrop of an Instagram post for a cover reveal (Prompt 13-12).

Prompt 13-12. Using Copilot to create a graphic for an Instagram post

I'm about to do a cover reveal on Instagram for a new fantasy book and I need a background. I will later put the cover image on top. But I need a background that looks cozy. Include fairy lights, flowers, and a knit blanket in the image, essentially framing the center, where the book will go. Again, do not add the book in yourself. Ensure that the center has enough blank space for me to include the cover image myself later.

Figure 13-12 shows the image ChatGPT generated in its response.

Figure 13-12. *A very cozy and magical backdrop of a cover reveal Instagram post.*
Disclaimer: Figure 13-12 is an AI-generated image

This is such a cozy-looking photo that would go great with a picture of a soft fantasy novel with a nice caption on an Instagram post.

Now let's look at how ChatGPT can generate images for graphic art!

Creating Graphic Art for Musical and Performing Artists

For indie musical and performing artists, resources can be hard to come by, especially when funds are low and there are other expenses to cover. This is why ChatGPT is a great tool for indie artists that simply need low-cost alternatives until they can afford illustrators or graphic designers that can craft a piece to fit their needs. So we'll be looking at how we can use DALL-E to create posters and album art covers.

Creating a Poster for a Flamenco Dance Concert Event

It's very exciting to be part of the process for creating graphics for a local event; whether it's hosted by the city or a small school, it's fun to just be part of the process. It's even more rewarding when you feel like you can offer a product that looks high in quality that cost these institutions that run on low funds very little. Let's give ChatGPT Prompt 13-13, where we explain our situation a little.

Prompt 13-13. Creating graphic art for a poster

I'm creating graphic art for an upcoming concert where there will be live flamenco dancers performing on stage. I need you to make a poster featuring a vibrant illustration of a flamenco dancer with a flower in her hair, dancing with a partner dressed in all black.

Figure 13-13 shows the image ChatGPT generated in its response.

Figure 13-13. *DALL-E makes a beautiful poster advertising a flamenco dance concert.*
Disclaimer: Figure 13-13 is an AI-generated image

This is such a dynamic image! And as you can see, the art style differs greatly from anything we've requested thus far. This is because we were so specific in our prompt about what this art would be used for, as well as the style we were looking for.

Speaking of different art styles, let's see what happens when we ask ChatGPT to depict more abstract concepts, like you might do for album cover art.

Creating an Album Art Cover for an Indie Rock Band

Indie rock is a genre that embraces odd aesthetics and out-of-the-box thinking. For this reason, the abstract concepts ChatGPT can generate are perfect for album covers. We used Prompt 13-14 to create a cover for an indie rock band called Red Horizon getting ready for their debut album entitled Genesis.

Prompt 13-14. Generating a cover for an indie rock band

```
We are an indie rock band called Red Horizon and we're coming
out with our debut album called Genesis. Genesis features
gentle rock music and lyrics that focus on coming out of
depression and likening the process to a seed sprouting from
the soil and emerging as a sprout. We need an album cover that
reflects this. We'd love floral details, but add a little
grunge. Because our band is called Red Horizon, we were
thinking about featuring roses and vines in the image, as well
as a color scheme that reflect colors of the dawn. We want the
album cover to have a lot of contrast and rough edges. We want
it to stand out. Generate this image.
```

Figure 13-14 shows the image ChatGPT generated in its response.

Figure 13-14. *A beautifully abstract and rough-looking album.*
Disclaimer: Figure 13-14 is an AI-generated image

This is a wonderfully expressive piece of artwork! Even if you weren't completely satisfied with the colors or wanted minor changes, this gives you a great starting point that you can edit any way you want.

So, let's try generating an album cover for a completely different genre and see how ChatGPT holds up.

Creating an Album Art Cover for a Jazz Band

While an indie rock band might look for ingenuity in a sometimes-messy way, we're going to try to create a cleaner design. Jazz is a genre that embraces and understands the ties between strong retro vibes and feelings of nostalgia. For this example, we're going to ask ChatGPT to create an album cover for a jazz band called Constance's Constellation for their album Harmony in the Routine. In our example, we're expressing to ChatGPT what the album aims to do and what we're looking for in cover art. We're being much more specific, as you can see in Prompt 13-15.

Prompt 13-15. Now we're generating an album cover for a jazz band

```
I'm Constance Laurence and I'm the leader of the jazz band
Constance's Constellation. We're coming up with an album called
Harmony in the Routine that we want to reflect the feeling of
hard working people that need inspiration for the backdrop
of tasks they do on a day-to-day basis. For that reason,
the songs on our album are high-tempo, rhythmic, and upbeat.
We want a cover that emphasizes simple things, so it should
feature an illustration of a vintage car. The image needs to
have a vibrant color scheme that uses white and yellow, with
blue highlights. The image should pop, but be simplistic and
minimal. Generate this album cover art.
```

Figure 13-15 shows the image ChatGPT generated in its response.

Figure 13-15. *DALL-E creates an album cover for a jazz band.*
Disclaimer: Figure 13-15 is an AI-generated image

As you can see, ChatGPT is capable of navigating different art styles when prompted to do so. Oftentimes, getting "wonky-looking" images from ChatGPT comes out of not knowing how to prompt it well or not being specific enough. But the best way to get better at anything is to practice. So get creative! Try different prompts and request different art styles (like cubism, vector, and pixelated; never ask for art in the style of living artists). Think out of the box and ask for different "camera" angles or expressions on your subjects' faces. The possibilities are endless.

Now, Let's Talk About ChatGPT's Sora Model

The images we've made so far have been with the help of ChatGPT's image generation model, DALL-E. However, ChatGPT also has a video generation model called Sora. Sora is able to generate short videos between 5 and 10 seconds long for anyone with a Plus membership and up to 20 seconds for those with Pro membership. Like with the image generation prompts we used, Sora can be prompted to generate surreal lifelike videos or even create a 3D animation style video as well.

Now, you might be wondering what you'd even do with such a short video. Even at 20 seconds, that's barely a TikTok video. Well, rather than thinking of creating something like, say, a movie or animated skit, think of allowing the video to *complement* other content. I'll explain what I mean by this, but first let's look at the interface for Sora, because it's different from everything we've seen so far.

Exploring the Sora Website

Now, while Sora is a ChatGPT model, it does have its own website, which can be accessed at `https://sora.com`. You'll log in with your ChatGPT Plus account, which we've done in Chapter 1. Once you're in, you should see the Sora home page, as shown in Figure 13-16, which will showcase all of the most popular videos created by the community, along with the prompts they used. I highly recommend scrolling through the home page for a while to get inspiration!

Figure 13-16. *Behold the Sora home page!*
Disclaimer: Figure 13-16 contains AI-generated images

Now, this might look a little overwhelming at first. But in reality, using Sora is almost identical to using ChatGPT. Let's briefly break down what we're looking at here.

A Numbered Guide to the Sora Home Page

As we've done several times already now, we've included a numbered guide to the Sora home page, as you can see in Figure 13-17.

Figure 13-17. *A numbered guide to the Sora home page. Disclaimer: Figure 13-17 contains AI-generated images*

Navigating the home page is very simple. As I said before, using it is very similar to using ChatGPT.

1. This is the text box that allows you to send a prompt to Sora and get a video in the response. More on this in a minute.

2. This is the explore feed that allows you to see recently created videos by the community, popular videos, and videos that you've hit "like" on.

3. This is your personal library where you can see all of your favorited videos, videos you've generated, and videos you've archived.

Now, the text box we're using to actually make the prompts has a few interesting settings I'd like to cover before we move on, so let's take a look at Figure 13-18.

I want a cozy, rustic aesthetic for a B-roll. Capture close-ups of a rolling pan rolling over dough and then flour being sifted in slow motion. |Use smooth pans, tilts, and zoom-ins for transitions between steps, and add time-lapse effects to show dough rising.

+   □ 16:9   ◈ 480p   ◷ 10s   ⊡ 2v   ?   Storyboard ↑

Figure 13-18. *Check out the Sora text box to look at your options for prompt settings*

At the bottom of the text box are a few settings that can be used to modify the quality and length of the video, as well as the size and dimensions. You can also allow Sora to generate two videos from the same prompt, which gives you the ability to choose which video you like best. It's up to you and the purposes you want your video to serve to determine how you want to adjust these settings. But be aware that you have a limited amount of credits you can use per month.

Ah, I'm sure you're now wondering what credits are. Allow me to explain.

What Are Credits, and What Are They Used For?

Depending on your subscription tier, you're allowed a certain number of "credits" per month. These credits represent how many videos you can create based upon the quality and length of the video. Shorter, lower resolution videos spend the least amount of credits. Short, high-quality videos spend a little more. A longer, low resolution video spends a little more than that, and a long, high resolution video will spend up the most.

ChatGPT Plus members get 1000 credits a month, which, according to their website, will equate to *about* 50 videos. But, it goes without saying that this number will be slightly lower or higher based upon the settings you use.

In my personal opinion, it's best to try out the shorter, lower resolution settings when you're just getting used to Sora. That way, when you're ready to use the software for actual products, you have a little more experience and will waste less credits on videos that look a little too wonky to use.

Believe me, in some of my first tries, I asked for an ice cream with chocolate syrup being poured over it. For some reason, I ended up with what I can only describe as an ice cream on a cone with a slight... balding problem (Figure 13-19).

Figure 13-19. *I still don't quite know how I ended up with this. Disclaimer: Figure 13-19 is an AI-generated image*

Alright, with all that out of the way, let's talk about *why* you would even want to make these very short videos.

Using Sora for Sleek and Modern Content Creation and Graphic Design

Nowadays, it's harder than ever to keep the attention of potential customers that are taking in most of their content through apps they're constantly swiping through. Getting them to click on links is hard enough. We also want to retain their attention and keep them from clicking off right away.

With this in mind, it's a good idea to use short videos in our content creation to maintain attention, whether that be on a website or portfolio landing page, or short clips in between YouTube videos; a little break from the monotony goes a long way in getting a viewer interested in paying attention to you.

So let's look at how we can make our content creation more interesting by using videos we make from Sora.

Using Sora to Create a B-Roll for a Cooking Video on YouTube

Do you know what A- and B-rolls are in film? Well, it's a simple concept. The A-roll features the main content of the video, whether it be a tutorial, an interview, or a person talking to a virtual audience on a platform like YouTube.

B-rolls are short clips that often show content related to the subject material of the video. Let's say the main video is an interview between the interviewer and a popular author. To break up parts of the conversation, the videographer might include footage of the author doing different things, like typing at their laptop, taking notes, or sitting at a cafe and people-watching for inspiration.

That said, you can see that although you currently can't use Sora to make longer content, it's certainly usable for scene breaks between footage. It can add a professional touch to otherwise monotonous videos. So let's try it out.

Crafting a Prompt for Sora to Create a Cozy Video of a Baker Kneading Dough

For our first example, we're going to be looking at how we can use Sora to create a B-roll for a YouTube video about making bread. We used Prompt 13-16.

Prompt 13-16. Giving Sora a detailed prompt to create a short video

```
A close-up shot captures hands gently kneading dough on a
rustic wooden table. The surface is lightly dusted with flour,
and the warm lighting casts a cozy glow, highlighting the
textures of both the dough and the grain of the wooden table.
```

You might have noticed that our prompt is very detailed. To get good results from Sora, it's best to be as descriptive as possible. After all, this isn't just an image we're talking about. We want to be specific. Where should the angle of the "camera" be? What does the lighting look like? What's happening in the video?

This might seem difficult at first, but try this: watch a YouTube video—it doesn't matter what the video is about; just pick one. Watch the first couple seconds of the video and imagine how you might describe it to someone that has blindfolds on. This kind of thinking will help you practice describing your concepts to Sora when it comes time to give it a prompt. Because, really, that's what you're doing. Sora clearly can't see what's in your mind's eye, and you have to be as descriptive as possible to get the result you're looking for.

It's Magic! Sora Generates an Uncanny Video of Someone Kneading Dough

Anyway, let's move on to see how Sora responds, as you'll see in Figure 13-20.

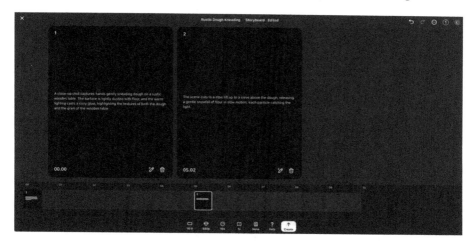

Figure 13-20. *Whoa, what's all this? What's storyboard mode?*

Note   If you're having trouble coming up with prompts, try describing what you want to create to ChatGPT, then ask it to give you a prompt for Sora!

Don't let this new screen scare you. It looks complicated, but the only thing that's happening is that Sora is giving you the ability to describe each scene of the video as it happens.

Now, I'm going to be honest. I didn't find much use in this feature. If the video you're making is only 5–10 seconds long, there really isn't much room for a transition between scenes. I feel as though this feature is more useful for those paying for the Pro version that can make 20-second videos. That's why for our purposes, we're going to skip using this feature.

When Sora enters this storyboard mode, it essentially fleshes out the prompt you gave it and shows you what kind of video it's going to make. After a few attempts, I decided that my prompt worked better, so I deleted Sora's suggestions and went with my original prompt.

The process of creating a video you're happy with is very similar to how we created images, meaning sometimes you can get a great result on the first try; other times, you're going to have to make modifications.

Figure 13-21 is a screenshot of the video I created with Sora that I liked best.

Figure 13-21. *Sora creates a video of a person kneading dough on a lightly floured surface.*
Disclaimer: Figure 13-21 is an AI-generated image

Now, if you notice, at the bottom of the screen is the truncated version of a longer prompt I'd used to critique what Sora produced the first few tries. However, in Figures 13-22 and 13-23, you can see the kind of prompts I used to modify the videos Sora generated.

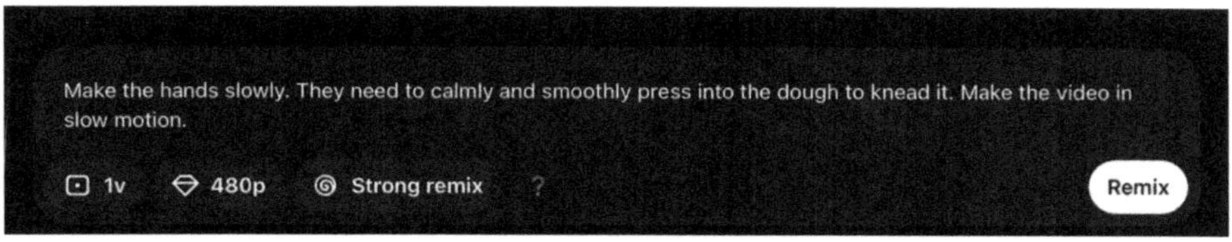

Make the hands slowly. They need to calmly and smoothly press into the dough to knead it. Make the video in slow motion.

⊡ 1v   ⬦ 480p   ⊚ Strong remix   ?   Remix

Figure 13-22. *Asking Sora to modify the video it generated of a person kneading dough*

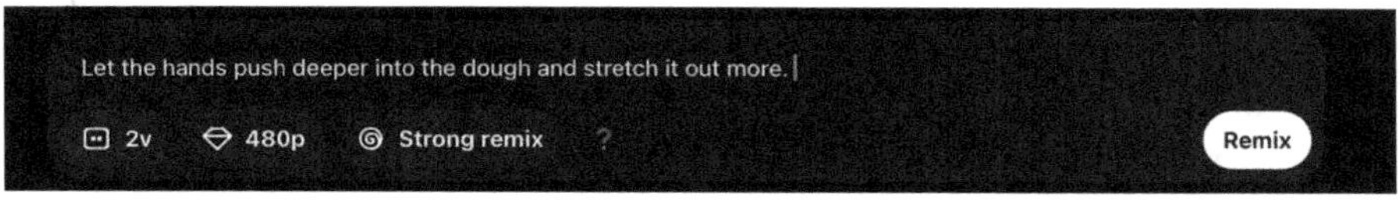

Let the hands push deeper into the dough and stretch it out more.|

⊡ 2v   ⬦ 480p   ⊚ Strong remix   ?   Remix

Figure 13-23. *Asking Sora to modify the video again*

Generating B-rolls for YouTube videos is just one way that videos made with Sora can be useful. You can also use it to make an animated background for a landing page or generate GIFs. You can add short clips to promotional videos or create a short concept to present to a videographer or graphic designer. Or if you're a graphic designer yourself, you can use Sora to draft concepts before you dedicate time to creating drafts on your own.

The possibilities abound. However, let's talk about what Sora *or* DALL-E should *not* be used for.

Using Sora and DALL-E Ethically

With the surging popularity of AI-generated content that comes closer and closer to producing lifelike results, an important question arises: How do we protect people from being harmed by this technology? Harm doesn't have to look like a robot takeover. Harm also comes in the form of impersonation, spreading misinformation, and replacing real people with something computer generated. These are real concerns, and thus it is up to us to be responsible for what we create when using these technologies. Therefore, we're going to be looking at some dos and don'ts when we're using DALL-E or Sora to generate content, though the same could be said for any content created using generative AI.

Don't Impersonate

This is imperatively important. **NEVER** impersonate a real person with the use of Sora or DALL-E. There are too many cases of the likeness of celebrities being used to generate content that can mislead the public into believing they did things that they didn't or using their likeness in ways they didn't approve of. This has happened to Taylor Swift, Steve Harvey, and Megan Thee Stallion, to name a few. I believe this to be highly unethical and would never support the generation of content that the people these images and videos are based upon did not consent to.

Don't Spread Misinformation

Just as videos and images of real people can harm the individuals being impersonated, videos and images meant to deceive the public into believing something that isn't true are extremely harmful and should be discouraged by all means. Insisting upon making people believe that something generated by AI is real is something I don't think any of us should tolerate to protect our spaces and the well-being of others.

Don't Hide the Fact You Used AI

I believe that many of the problems people are facing with AI content is the fact that it can be blended in with human-generated content, and one can be confused for the other. For this reason, I think it's extremely important to be transparent about when AI is being used to generate content, whether it be written or visual, so that we're all giving proper credit where credit is due. Transparency and accountability go hand in hand, and when we're accountable for what we create with generative AI, we can create a safer environment for everyone.

Do Generate Original Content

When generating content that features fictional people (and not recreations of real people), describe the kind of person you want in the depiction. Be as descriptive as possible! Again, I would encourage you to imagine you're describing something you can see to a person with blindfolds on. This allows you to generate original content that you can use for concept art, collaborating on ideas with team members, and personal inspiration.

Do Support Artists and Content Creators

As stated before, generating content with AI should never be used to replace an actual human being, and I would argue that it's actually near impossible to do so successfully. None of the prompts used in this book could be used to replace a person's creativity, and all of them required the experience and ingenuity of a person that knows what they're doing to create. And in the end, humans need to be the ones regulating the content being created. Humans should edit generated text, adjust generated images, scrap ideas, and go with others.

I think that AI is best used as a collaborative tool. Share ideas with someone you're working with by generating an image right quick or draft up an outline for an article with a few prompts and go over the concept with your editor.

Likewise, these tools are best utilized when collaborating with creative people that can bring these AI-generated ideas to life. In my experience, the best person to create generated art is an artist. The best person that can use ChatGPT to code is a programmer. The best person to use generated text for writing is an author.

Human art and creativity should always be valued over all else, and AI should never be an excuse to disrespect the people it should be empowering.

Do Be Honest About Using AI

Again, being open about the technologies we're using to create content is all the more important the more these technologies develop. We need to develop our ethics along with it, and we can only achieve that by indicating when generative AI has been used to avoid misleading information. You become someone that others can trust when you're open about your creative process, and therefore nothing you create can make other people feel as though they've been lied to.

This is why all content generated by ChatGPT in this book has been indicated as so. You should never attempt to deceive others through the use of AI by any means.

Conclusion

In this chapter, we've looked at the different capabilities of ChatGPT when we ask it to create images. We looked at all of the different ways we can prompt it and critique the results to get what we're looking for. We also looked at the ethical implications of AI generation and AI technology as a whole. We considered our responsibilities as people to protect the safety of ourselves and others while making use of new and exciting technologies at the same time.

Index

A

AI, *see* Artificial intelligence (AI)
American constitution, 287–295
Artificial intelligence (AI)
    prompt, 7
    TikTok, 5

B

Budget planning
    grocery receipt, 329–332
    meaning, 327
    o1 model, 328, 329
    spending details, 333

C

Canvas model, 3
    chat window, 113–116
    conceptualization, 116–121
    direct instructions, 129–133
    feature, 108, 109, 111, 112
    highlights, 113
    outline development, 124–130
    prompt approach, 133–139
    switching model, 110
    Tolkien, J.R.R., 120–123
ChatGPT, 1

account creation, 9
    chat window, 11, 12
    sign-up page, 9, 10
    upgrade plan button, 13
accurate results, 4
AI (*see* Artificial
    intelligence (AI))
boosting productivity, 4
budget (*see* Budget planning)
Canvas feature, 3
conversation role, 24–30
image generation, 3
lease agreement (*see* Lease
    documents)
mobile app (*see* User interface)
nondeterminism, 7, 8
prompt (*see* Prompt
    engineering)
teaching style (*see* Teaching
    ChatGPT)
verbal conversation, 2
word document, 2
Content creation, 39
    beauty vloggers, 48
    chat window, 44–47
    cite sources, 58–60
    marketing strategy, 41–43
    meal plans/stressing, 61

© Lydia Evelyn 2025
L. Evelyn, *Making ChatGPT Work for You*, https://doi.org/10.1007/979-8-8688-1445-7

U, V

W, X, Y, Z

GPSR Compliance
The European Union's (EU) General Product Safety Regulation (GPSR) is a set
of rules that requires consumer products to be safe and our obligations to
ensure this.

If you have any concerns about our products, you can contact us on

ProductSafety@springernature.com

In case Publisher is established outside the EU, the EU authorized
representative is:

Springer Nature Customer Service Center GmbH
Europaplatz 3
69115 Heidelberg, Germany